D1629869

DATA SOLILOQUIES

Richard Hamblyn &
Martin John Callanan

Data Soliloquies

Richard Hamblyn & Martin John Callanan

Published in London at the Slade Press by the Slade School of Fine Art
for *UCL Environment Institute*, University College London,
Gower Street, London, WC1E 6BT. United Kingdom

A CIP catalogue record for this title is available from the British Library.

ISBN 978-0903305044

Design by Slade Print
Typeset in Garamond, Dagny and Museo Sans
Printed by Gutenberg, Malta

DATA SOLILOQUIES

Richard Hamblyn &
Martin John Callanan

UCL Environment Institute
London
2009

FOREWORD

Professor Mark Maslin
Director of the UCL Environment Institute

Writing a foreword is a tricky thing, because you can never be sure when (or if) the reader will look at it. Will they read it before they start the book, or only after they have finished the rest? So I thought I would take the safe route and describe the background to the incredible creative partnership between Richard and Martin, the co-creators of this book.

It all started one misty morning in Cornwall while I was on holiday: a simple phone call asking if I would take over the Environment Institute at University College London. When I realised that the Provost intended to provide serious money to make it work, I of course said 'yes', and started to plan my dream team. The idea behind the UCL Environment Institute is to bring together research across all disciplines within the university to tackle important environmental questions ranging from climate change to the causes of evolution, from waste management to designing sustainable cities. The Institute is also there to shout from the roof-tops how brilliant the work on environmental issues is across the university. This is even easier now the *Times Higher Education Supplement* has judged UCL to be the fourth best university in the world, renowned for its work on the environment and on global health.

From the outset, the Environment Institute was representative of nearly every disciplinary area, apart from the arts and humanities. Through my work as a trustee of the charity TippingPoint, I was aware of the growing interest and engagement in climate change and the environment across the arts community; but how could I persuade artists and writers to become involved in the work of the Environment Institute? But then something happened that could never have been planned. A brilliant ex-student of mine, Keith Leung, contacted me and asked if UCL would be interested in starting a creative partnership with his company JLT Re, a major reinsurance company.

As part of this collaboration JLT Re offered a generous three-year stipend that allowed me to start something new and exciting – namely, an artist and writer in residency programme at the Environment Institute. But where to start, since I must admit that my knowledge of art and literature starts with my children's drawings and ends with trashy paperback novels. So we asked the UCL Slade School of Fine Art and the English Department if they would co-host these positions and help decide who we should appoint. I also twisted the arm of a good friend of mine, Peter Gingold (the Director of TippingPoint), to give us a third view. We were blown away with the response: more than 100 applications in the first year; after a lot of heart searching and interviews we appointed Richard as the writer in residence and Martin as the artist in residence. And, oh did we fall on our feet. Not only have these two been individually brilliant, but by complete chance there is this amazing creative energy that occurs whenever they are in the room together. This book is a testament to both their individual and collective brilliance.

What makes *Data Soliloquies* such a brilliant and unusual book is its creative response to the ways in which climate change is imagined by the wider culture. The public understanding of science is not only a growing academic research area, it is also of crucial importance to wider society, helping us to understand the ways in which we apprehend the world through metaphor-laden imagery. The book tackles the empowered 'words', 'numbers' and 'diagrams' of climate change, the emotive and political sound-bites, and the visualization of scientific and publicly generated data. By offering insights into the sometimes alarming public career of scientific data, this thoughtful book makes a potent contribution to the public understanding of 'the public understanding of science'. I hope you enjoy it as much as I did*.

*This book, however, is not the end of the story but rather a beginning. Firstly, Richard and Martin will continue to work together into the future and we await their next creations with bated breath. Meanwhile, the residency programme goes from strength to strength, with over 180 applications this year, from which we have selected two equally brilliant people: Jean McNeil, an insightful and world-renowned novelist whose latest work is *The Ice Lovers*, which I thoroughly recommend; and Subathra Subramanyam, an inspiring contemporary Indian dancer and choreographer, who is also a trained science teacher and has led two Cape Farewell Youth Expeditions to the Arctic in order to study both the science and art of climate change.

I never said it. Honest. Oh, I said there are maybe 100 billion galaxies and 10 billion trillion stars. It's hard to talk about the Cosmos without using big numbers — but I never said "billions and billions." For one thing, it's too imprecise. How many billions are "billions and billions"? A few billion? Twenty billion? A hundred billion? For a while, out of childish pique, I wouldn't utter or write the phrase, even when asked to. But I've gotten over that. So, for the record, here goes: "Billions and billions."

Carl Sagan, 1997

ONE HUNDRED AND TWENTY-FIVE BILLION METAPHORS PER SECOND

This publication grew out of an intermittent collaboration that began during our terms as writer and artist in residence at the UCL Environment Institute for the 2008-09 academic year. Throughout that time we maintained an ongoing conversation that centred on a mutual interest in the use and abuse of scientific data, particularly in its visual manifestations. A wide array of graphs, charts, computer models, diagrams and other forms of visual advocacy have become inescapable fixtures of public science presentations — particularly in the field of climate science — though they are often treated as if they were neutral 'found objects' rather than elaborate narrative constructions containing high degrees of statistical uncertainty. These fascinating, and occasionally beautiful, artefacts soon became our shared subject, and much of the work that we produced during our residencies, such as (RH's) 'The whistleblower and the canary: rhetorical constructions of climate change', and (MJC's) *Text Trends* animations, dealt explicitly with the theatricality of data display and the spectacularization of scientific information. 'What is so special about the language of quantity?', as the statistical historian Theodore Porter asked at the outset of his seminal study *Trust in Numbers* (1995), and much of our collaborative work was undertaken in response to this and other questions concerning the extraordinary cultural fluidity of scientific data.

Our title, 'Data Soliloquies' — the term was coined by Jon Adams, of the London School of Economics, as a description of the graphics-laden science 'fact' novels of the late Michael Crichton: notably *Prey* (2002); *State of Fear* (2004); and *Next* (2006) — reflects the way in which scientific graphs and images often have powerful stories to tell, carrying much in the way of overt and implied narrative content; but also that these stories or narratives are rarely interrupted or interrogated. They are information monologues — data soliloquies — displayed more for their visual and rhetorical eloquence than for their complex (and usually hard-won) analytical content. As a case in point: the best known scene from Al Gore's Oscar-winning *An Inconvenient Truth* (2006), when Gore was hoisted by hydraulic lift some ten metres up the y-axis of an elongated graph that plotted likely atmospheric concentrations of carbon dioxide, projected to the year 2100. After 2050, when a projected concentration of some 550 parts per million had been reached, the line went vertical, soaring beyond the furthest extent of the fireman's lift, in a dramatic literalization of the statistical concept of a measurement going off the scale. 'You've heard of 'off the charts?", asked the elevated Gore, to a ripple of nervous laughter from the audience.

Our previous choice of title, now demoted to a chapter heading (above), was the more cumbersome — and quickly outdated — 'One Hundred and Twenty-Five Billion Metaphors Per Second', which was also a reference to uninterrupted data, in this case the workload capacity of the old (2004-09) Met Office supercomputers at the Exeter campus, which were capable of processing around 125 billion pieces of weather information per second. Given the artful ways in which weather data is modeled and visualized for public display, this amounted, we reckoned, to the creation of around 125 billion metaphors per second.

That earlier title, with its conscious echo of Carl Sagan's notorious 'billions and billions' tic, was overturned in the course of one of the regular email exchanges that tended to follow our spoken conversations, exchanges in which we amended or refined some of the points that arose, and which also formed an ongoing record of what was said and unsaid over the course of the year:

Martin: The 'digital revolution' has been both powered and measured by the ability to transmit, process and store information. PCs become out of date within two years of manufacture because under Moore's Law, the power of chips to process information doubles every 18 months. According to research published in November 2003 by the University of California, more information has been created and stored in the last five years than at any time before in human history. In 2002 print, film, magnetic and optical storage media produced about five exabytes of new information. One exabyte is a billion gigabytes: somewhat over a quintillion (or ten to the eighteenth power) bytes. Five exabytes of information is the equivalent of half a million new libraries the size of the print section of the U.S. Library of Congress, which is America's library of record. They believe that at this current rate, 800MB of information is produced for each member of the human race each year. That would take over nine meters of books to store. Remember this is new information. It is already double what was happening three years before that; and these figures are now more than six years old: computer access has been on the rise and the network is forever expanding. The world's fastest computer in 2006, IBM's Blue Gene L, has more processing capability than the 500 most powerful computers of 2001 combined. Blue Gene L is 15 times more powerful than its predecessor: within five seconds it can produce a volume of data equivalent to the total information held in the British Library. The data collected by our networks, in data warehouses and elsewhere, vastly exceeds that which could be recorded about our world and knowledge on the 1:1 scale imagined by Borges.* And today these numbers are small in comparison.

Richard: Coincidentally, the Met Office is just about to unveil its new £33 million IBM supercomputer: it will be switched on at the end of May 2009, but will take two months to boot up, and won't commence forecasting until August 2009 at the earliest; and it won't reach peak performance until 2011, after the first of its 'midlife upgrades'. The Met Office specialist who negotiated the contract with IBM, Peter Williams (Met Office High Performance Computing Programme Manager, to give him his full title), spent three years designing the specifications, which make interesting reading:

- after 2011 it will be capable of processing nearly 125 *trillion* calculations per second (well, there goes our title: 125 billion seems a bit trifling by comparison; and 'trillions' just doesn't have that *Carl Sagan* bathos);

- with 15 million megabytes of memory, and operating at close to a petaflop, it will be more powerful than 100,000 standard PCs combined;

- it will occupy two halls, each the size of a football pitch, and require 1.2 megawatts of energy to run (enough to power a small town);

- but it will be a midget compared to the Sequoia system, the 20-petaflop supercomputer that is currently being constructed at the Lawrence Livermore National Laboratory in California, on behalf of the United States Department of Energy, who will use it to simulate nuclear blasts. (Now there's an interesting cultural comparison: Britain's most powerful supercomputer is devoted to modeling weather; America's most powerful supercomputer is devoted to modeling weapons.)

* Borges's fictional map, described in 'On Exactitude in Science', imagines an empire where the science of cartography has become so exacting that only a map on the same scale as the empire itself is sufficient. See p. 21.

I remember visiting what I assume is now the Met Office's ex-computing suite, just after they moved to the current site in Exeter, and the place was amazing — you had to wear anti-static overshoes in case of electronic interference, and not move faster than a stroll through the basement space, which was the size of a college sports stadium. It was filled with the disparate elements of the combined NEC SX6 and SX8 systems, which looked like oversized vending machines, all flashing away with coloured lights, and bundles of wires spilling across the floor. The noise was extraordinary, with dozens of powerful air handlers roaring away, and it struck me as strange that the smaller and faster processors become, the bigger the boxes they come in, as rapid air cooling and circulation become ever more important. Nevertheless, it was hot as anything in there.

But now all that five-year-old equipment — which had itself replaced the so-called 'Cray Twins', the T3E Massive Parallel Processors installed at Bracknell in 1997 (at the time the third most powerful computer in the world) — has been ditched for the new one-petaflop IBM supermachine, which will be a thousand times faster than the previous model, and 'a million, million times faster' than the first computers used by the Met Office in the 1950s, which delivered a mere 1,000 calculations per second. The 2004 NEC machines were capable of processing around 125 billion pieces of information per second (being thirteen times more powerful than the 1997 T3Es), but this new 2009 IBM model, which still doesn't have a name, will soon be able to process around 125 trillion calculations per second, based on data gathered from thousands of weather stations, balloons, satellites and other atmospheric observations all around the world. That's the equivalent of nearly 200,000 computations *per second* for every human being on the planet. Your telling point, that now, every second, more information is created than has ever existed before, conjures a kind of mad, mythic enterprise, more Borgesian than anything imagined by Borges, made dizzyingly 'real' by vast and unfathomable machines such as these.

I came across this neat chart the other day, showing an updated classification of processing power. Flops ('floating point operations per second') are a measure of a computer's performance capacity, especially in the fields of scientific computing, which make heavy use of floating point calculations; *flops* are essentially a refinement of the old Instructions Per Second (IPS) unit, with the same Greek-language SI prefixes ('mega', 'giga', 'zetta', 'yotta') rising by a factor of a thousand each time:

Name	Operations per Second
yottaflops	one septillion (10^{24})
zettaflops	one sextillion (10^{21})
exaflops	one quintillion (10^{18})
petaflops	one quadrillion (10^{15})
teraflops	one trillion (10^{12})
gigaflops	one billion (10^{9})
megaflops	one million (10^{6})
kiloflops	one thousand (10^{3})

So far no computer on earth is as powerful as an exaflop, though it is only a matter of time — and it's anyone's guess when the world's first yottaflop computer will appear.

Martin: In 1968 Glenn Gould said: 'I think technology exercises a charity; one which we have not yet reflected. There is a real charity in the machine, because it is there to help man, but of course it can be perverted. In that way we are not very free among our creations. But, in itself, it is good. It is the network of machines and techniques that encompass the earth. It is the ensemble of all the networks, radio networks, television network, oil network, hydraulic network, railroad network, telephone and telegraph, and all of that. So that it is impossible to consider a machine as isolated from the rest, it is part of all the rest. So that there is only one machine in fact encompassing the earth. And that has a meaning. And that machine steaming from the activity of man is between man and nature like a second nature; offering to us its mediation. We cannot go to nature now without going though the network...'

Richard: That reminds me of how, during the course of my Met Office visit, it became apparent that state-level meteorology can no longer be considered a branch of physics (which is what meteorology slowly became over the course of the twentieth century); it is, instead, a function of information technology. Today's Met Office is little more than a vast information plant, with data streaming in and out at inconceivable volumes. But this has always been the central feature of state-level weather science: right from the start, when the British Board of Trade established its weather bureau in the 1850s, under the directorship of Robert FitzRoy, the main challenge was what to do with the overwhelming quantity of local weather information that poured into the central department day after day. The statistical avalanche soon proved too burdensome to process, and during its first few decades the Meteorological Office, as it began to be called, had to content itself with fulfilling a largely archival, rather than its intended predictive, function. The overload became so severe that when Sir Napier Shaw took over as director of the Met Office in 1907, he was quoted as saying that 'the best thing for meteorology would be for everyone to stop observing for 5 years.'[1]

[1] Cited in Katharine Anderson, *Predicting the Weather: Victorians and the Science of Meteorology* (Chicago: Chicago University Press, 2005), 12.

But even then, a hundred years ago, information overload was nothing new — eighteenth and nineteenth-century reviewers had already complained in vain about the backlog of new books and journals that kept appearing week after week ('our presses groan with new publications on every subject under the sun'); and by the mid-nineteenth century it was already impossible for every new work to be reviewed. It's an interesting corner of intellectual history, this; in fact I'd love to write a cultural history of information overload — and in so doing, of course, add a little extra to the pile.

In that Empire, the craft of Cartography attained such Perfection that the Map of a Single province covered the space of an entire City, and the Map of the Empire itself an entire Province. In the course of Time, these Extensive maps were found somehow wanting, and so the College of Cartographers evolved a Map of the Empire that was of the same Scale as the Empire and that coincided with it point for point. Less attentive to the Study of Cartography, succeeding Generations came to judge a map of such Magnitude cumbersome, and, not without Irreverence, they abandoned it to the Rigours of sun and Rain. In the western Deserts, tattered Fragments of the Map are still to be found, sheltering an occasional Beast or beggar; in the whole Nation, no other relic is left of the Discipline of Geography.

Jorge Luis Borges, 1946

In May 2006 the television naturalist Sir David Attenborough — the most trusted voice in British broadcasting — announced that he was no longer a sceptic when it came to the causes of climate change: 'My message is that the world is warming, and that it's our fault', he informed a startled looking Huw Edwards on the BBC's *Ten O'Clock News.* But it hadn't been images of glacial retreat that had served to convince him of humanity's guilt, nor had it been footage of Arctic icecaps crashing into an ever-rising sea. His conversion had been brought about by a small coloured graph produced by researchers at the Hadley Centre, the climate change wing of the Met Office. The graph featured three jagged lines, one red, one green and one yellow, representing, respectively, average recorded temperatures, natural climatic variability, and atmospheric concentrations of carbon dioxide, each plotted over a 150-year period (from 1850 to 2000). Up until the middle of the twentieth century, the three lines rose and fell together, but after that point, while the green line rose only very slightly, the red and yellow lines shot up in tandem, J-curving dramatically from the 1970s onwards. As Attenborough pointed out, 'the coincidence of the curves made it perfectly clear that we have left the period of natural climatic oscillation behind', and that our climate is now, effectively, man-made (fig. 1).

Such graphical representations have come to assume a key role in climate change debates, often presented in the manner of exhibits at a trial, and credited with the irrefutability of

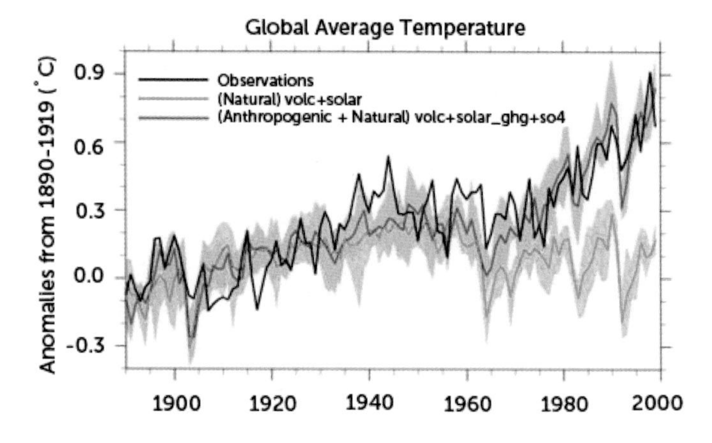

Fig. 1
the Parallel Climate Model that convinced Sir David Attenborough

unmediated data, particularly by non-scientists, who rarely appreciate the creative input that goes into making these complex visual statements. This chapter will consider the design elements of a handful of examples of scientific data display, with a particular focus on some of the best known climate science graphics from the past fifty years.

THE KEELING CURVE

Let's start with the Keeling curve, arguably the most influential scientific graph ever made (fig. 2).

Ever since Charles David Keeling of Scripps Institute of Oceanography began to collect atmospheric carbon dioxide (CO_2) data at the Mauna Loa observatory during the International Geophysical Year of 1958, the story of global climate change has been narrated through a series of upward-trending graphs — Keeling's being the first and easily the best known of the sequence, not least because of its starring role in Al Gore's graphics-heavy PowerPoint documentary, *An Inconvenient Truth* (2006).

The visual effectiveness and explanatory power of Keeling's design is due to its uncluttered simplicity. Its content is

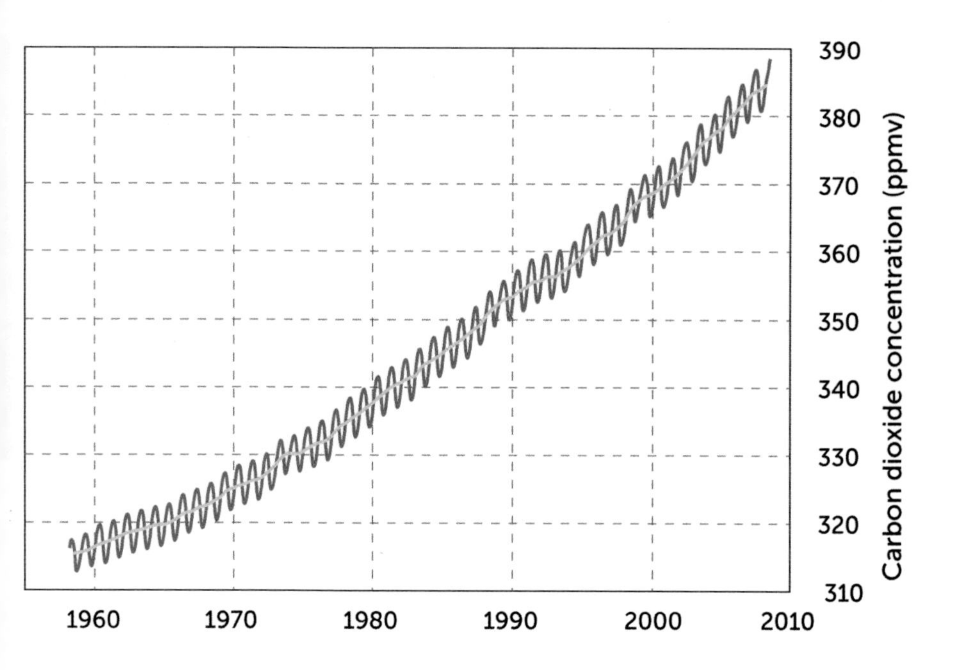

Fig. 2
The Keeling curve, 1958 to now

easily readable as a year-on-year increase in carbon dioxide concentrations from around 315 parts per million by volume (ppmv) in 1958 to its current (late 2009) levels of 390 ppmv; and then there its distinctive shape, the 'fearsomeness' of the (almost) steadily rising red line being much enhanced by its saw-toothed jaggedness, the regular up-and-down variations caused by the annual greening and wintering of the northern hemisphere — a seasonal leafy uptake of carbon dioxide that serves to reduce the ever-increasing atmospheric load by around 5 ppmv every year. The jagged Keeling curve is probably the most important data set in environmental science, and has become something of a freestanding scientific icon in its own right, as can be seen in fig. 3, which features the plaque from the Keeling Building at Mauna Loa, and a reconstructed still from *An Inconvenient Truth*, showing the moment when the entire screen was briefly filled with a de-contextualized close-up of the blood-red saw-stroke, a potent visual synecdoche for the inexorability of human-enhanced climate change.

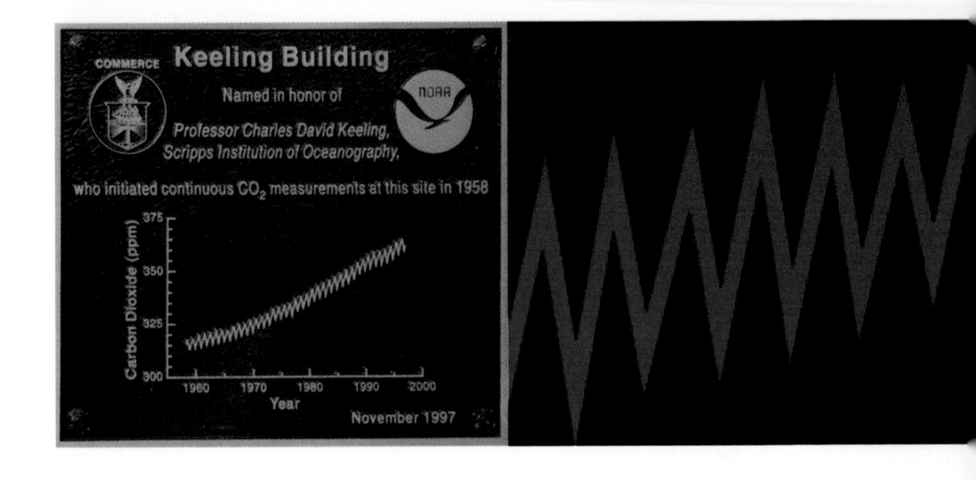

Fig. 3
The Keeling curve as a freestanding icon: The Keeling Building, Mauna Loa, Hawaii; screengrab from *An Inconvenient Truth* (2006)

THE HOCKEY STICK CONTROVERSY

The explanatory elegance of the Keeling curve has often been emulated by other climate scientists, but not always with the same degree of success. Michael Mann *et al*'s 'hockey stick' curve, for example, was designed to show reconstructed estimates of Northern Hemisphere mean temperature changes over the past 1,000 years (fig. 4).[2]

The term 'hockey stick' was not Mann's own (Mann preferred 'the MBH98 Reconstruction'), but the hockey-stick moniker was widely adopted during the dispute that followed the graph's appearance as a key piece of evidence in the Third Assessment Report of the Intergovernmental Panel on Climate Change (published in 2001). The dispute centred on technical aspects of the methodology and data sets used in creating the reconstruction, particularly the reliance on non-instrumental proxy data such as tree-rings, ice-cores and isotopic analyses of coral. Looked at from an information design perspective, however, the graph is not all that easy to follow. For a start, it does not show actual temperature rises over time, but 'departures from the 1961-1990 average', as represented by the horizontal bar at degree 0 — a somewhat confusing choice of 'start'

[2] From Michael Mann *et al*, 'Global-scale temperature patterns and climate forcing over the past six centuries', *Nature* 392 (1998), 779-87.

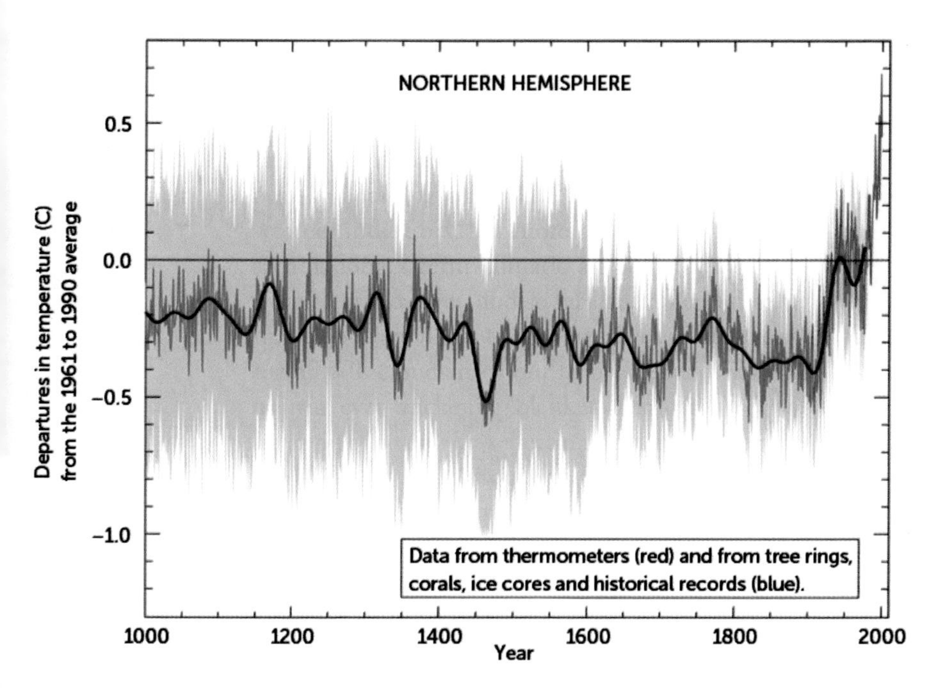

Fig. 4
Michael Mann *et al*'s 'hockey stick' graph

temperature. And visually there is a lot going on, particularly with the light grey shading in the background, which indicates uncertainty ranges: literally 'the grey area'.

Mann's visualized data attracted a lot of criticism for the way that it appeared to emphasize the flatness of the 900-year average line (the hockey stick's 'shaft'), which then J-curves abruptly from the turn of the twentieth century to form the dramatically rising 'blade'. Suggestions were made that Mann and his team selected or even modified their proxy data in order to create a smoother historical lead-up — thereby giving a misleading visual impression that recent rates of warming are unprecedented. After all, the grey shading of the 'uncertainty' region shows far greater temperature anomalies than appear in the more recent instrumental data (coloured red). Mann's critics, such as the climate scientist Hans von Storch, who is far from sceptical about the reality of anthropogenic warming, have pointed out that the graph contains a range of assumptions that are simply not statistically permissible, even though they agree that the data does show an overall temperature rise occurring over recent decades.[3]

These and other objections prompted a lengthy correspondence between Mann and his critics, which culminated in a Congress-appointed National Research Council report, published in 2006, which concluded that any statistical shortcomings in Mann's analysis were too minor to have influenced his overall findings.[4] It is worth reiterating that the NRC found in favour of the hockey stick's credibility, as it is often claimed (incorrectly) on Internet forums that the graph has been discredited as 'junk science'. That is not so, and as Hans von Storch recently commented, 'looking back we are satisfied about what has been achieved — namely an open-minded and exciting discussion about the merits and problems related to different methods; an atmosphere where the mere claims about the informational content of proxy-data meet a more critical response.'[5]

But from our perspective, the hockey stick controversy — which served as a boisterous political distraction from the underlying issues of anthropogenic warming — should not have focused solely on the disputed objectivity of the data (which we, of course, are in no position to judge), but also on the design of the graph itself. For the 'hockey stick', viewed as a purely visual artefact, was remarkably poorly constructed, particularly in its use of the 'zero-degree' line, the hesitant way in which the uncertainty ranges were represented, and especially the way that the *flatness* of the 900-year lead-in was so clumsily insisted upon, as though the shape of an averaged historical curve makes the slightest difference to how the world should act on its changing climate *now*. As von Storch aptly commented, the dispute in the end was all about 'the wobbliness of the shaft' — which is to say, it was all about the way the data was visually displayed. Had closer attention been paid to the problems of quantitative display, the result might well have been a less 'iconic' graph (compared to Keeling's classic curve), but it might also have avoided a lot of needless trouble.

[3] Hans von Storch *et al*, 'Reconstructing past climate from noisy data', *Science* 306 (2004), 679-82.

[4] *Surface Temperature Reconstructions for the Past 2000 Years* (Washington, D.C.: National Academies Press, 2006).

[5] Hans von Storch and Eduardo Zorita, 'The decay of the hockey stick', *Climate Feedback*, 3 May 2007: http://blogs.nature.com/climatefeedback/2007/05/the_decay_of_the_hockey_stick.

Such design-centred criticisms of data display have been the life's work of the information designer and statistician Edward Tufte, whose sequence of polemical (and visually stunning) publications includes *The Visual Display of Quantitative Information* (1983); *Visual Explanations* (1997); and, more recently, *Beautiful Evidence* (2006). His subject is the communication of complexity through design, and his claim that 'a lack of visual clarity in arranging evidence is a sign of a lack of intellectual clarity in reasoning about evidence', has become something of an axiom among statistical designers and thinkers. His core conviction — that the mind needs to access the visual in order to think clearly and well — has shaped his most original argument, which is that visual displays in science are important 'not just because they confirm or disconfirm testable theories, but because they generate knowledge that would be unavailable in any other form':

> Reliable knowledge grows from evidence that is collated, analyzed, and displayed with some good comparisons in view. And why should we fail to be rigorous about evidence and its presentation, just because the evidence is part of a public dialogue, and is meant for the news media? For information displays, design reasoning must correspond to scientific reasoning. Clear and precise *seeing* becomes as one with clear and precise *thinking*.[6]

[6] Edward R. Tufte, *Visual Explanations: Images and Quantities, Evidence and Narrative* (Cheshire, Conn.: Graphics Press, 1997), 52-53. See also John Grady, 'Edward Tufte and the Promise of a Visual Social Science', in Pauwels (ed.), *Visual Cultures of Science* (2006), 223.

Tufte has often written about the politics of scientific data displays, and one of his best known perorations concerned the so-called 'hyped Venus' affair, which centred on the 22.5-fold exaggeration of scale in NASA's widely circulated Magellan Project images from the early 1990s (figs. 5 & 6). The Magellan space probe circled Venus several times during 1991-93, collecting radar images of the planet's surface through its opaque atmosphere; these were then digitally remastered to form a colourful rollercoaster video tour of deep vertiginous valleys and canyons. As Tufte pointed out,

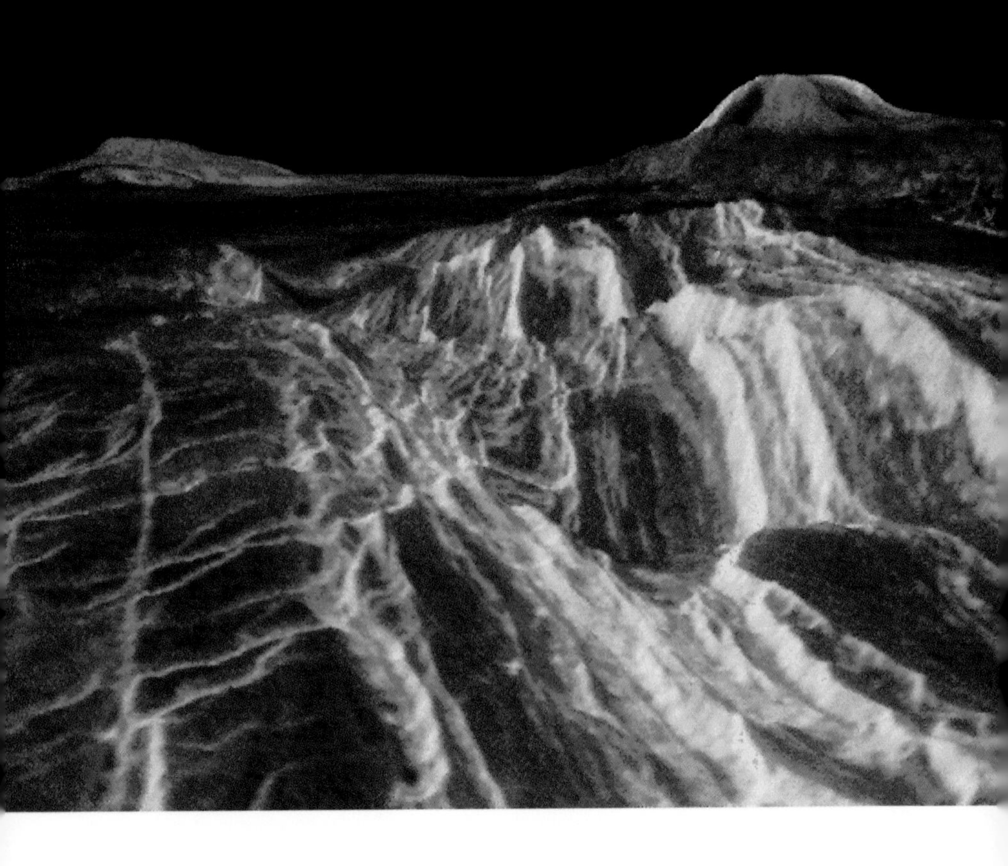

Figs. 5 & 6 'hyped Venus' from NASA's Magellan mission, 1990-94

the 'extravagent dequantification' of the computer-enhanced graphics made it look as though the planet was formed of vast mountain ranges rather than the flat volcanic plains of reality; all in all, it made for 'terrific television, but lousy science', and the episode briefly jeopardized NASA's reputation as a source of reliable data.[7] Jon Lomberg, the artist and designer who worked with Carl Sagan on the *Cosmos* television series, as well as on numerous NASA projects, has said how much it troubles him that a generation of viewers have grown up with the idea that the universe is some kind of day-glow theme-park, and such was the disquiet among planetary researchers in the wake of the Magellan debacle, that the astronomer David Morrison called for the founding of the Flat Venus Society in protest at NASA's distortions.

[7] Tufte, *Visual Explanations*, 23-4.

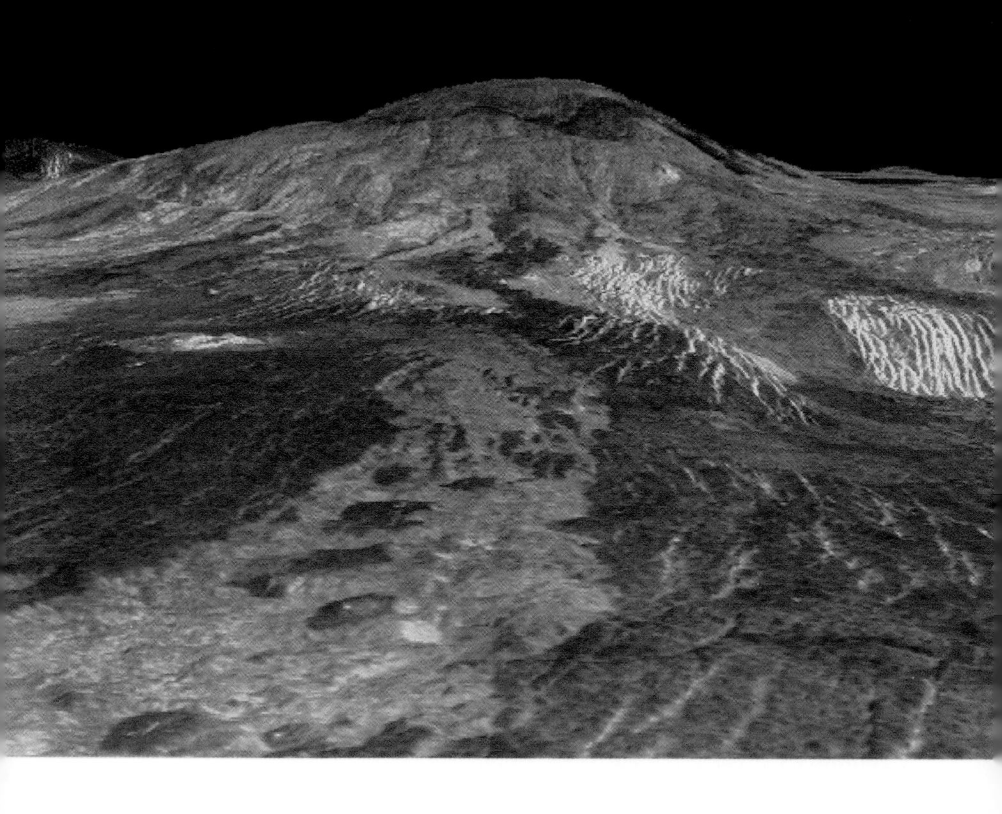

As Morrison famously complained, the slopes of Venus might be largely flat and featureless, but 'the public thinks they are precipitous peaks with near-vertical walls rising into a black sky. A *black sky*? On *Venus*?'[8]

[8] David Morrison, 'Forum: Flat Venus Society Organizes', *EOS* 73:9 (1992), 99. Clifford Stoll, in his entertaining (though dated) polemic, *Silicon Snake Oil: Second Thoughts on the Information Highway* (New York: Doubleday, 1995), also lambastes NASA's 'cartoon volcanoes'.

According to Lomberg, NASA also has a credibility problem when it comes to colour imaging, there being several colour categories to choose from: 'true colour', which comes close to what might actually be seen from space; 'false true colour', which offers an intensified version of true colour, allowing features such as clouds to show up more clearly; and 'true false colour', in which different true colours are contrasted and combined to produce an entirely new spectrum (again to allow faint details to be shown). But then there is another category altogether, used for images such as the famous astrophotograph taken from the Hubble Space Telescope in 1995, and dubbed 'The Pillars of Creation' — three vast columns of dust and gas in which new stars are forming — which Lomberg describes as 'not even false true colour (where the real colours are intensified). Totally arbitrary colours; they've put a green background in, which makes a very nice contrast with the reddish colour of the pillars; and it's great to see the shapes and see the morphology, and it seems very 3-D and, like the best nature photographs, it has a powerful emotional content. But the colours have nothing to do with reality.'[9]

[9] Jon Lomberg, interviewed on 'In Conversation', Radio National (Australia), 5 January 2006. Transcript available at: http://www.abc.net.au/rn/science/incon/stories/s1536397.htm

GRAPHIC DECEPTION

But there are far more serious cases of graphic irregularity than NASA's digitized excursions into sub-space-age science fiction. When the Goddard Institute for Space Studies at Columbia University, New York, changed the layout of its website in 2005, a number of its historical climate change graphics were replaced with redrawn examples that were apparently designed to emphasize the scale of recent warming. Figure 7 shows a before-and-after example of one of these redrawn graphs.

The earlier graphic on the left shows annual mean temperature data from Stuttgart for the period 1750-2000, with significant peaks appearing in the 1840s and 1850s. Note that the y-axis goes up to 12 degrees C. From this graph, it can be seen that recent warming in Stuttgart appears to be lower than it has been at other times over the past two centuries. This itself is not surprising — climate warming figures tend to be globally averaged, so it is to be expected that some places will show localized cooling, some will show localized warming, while others will show no change at all.

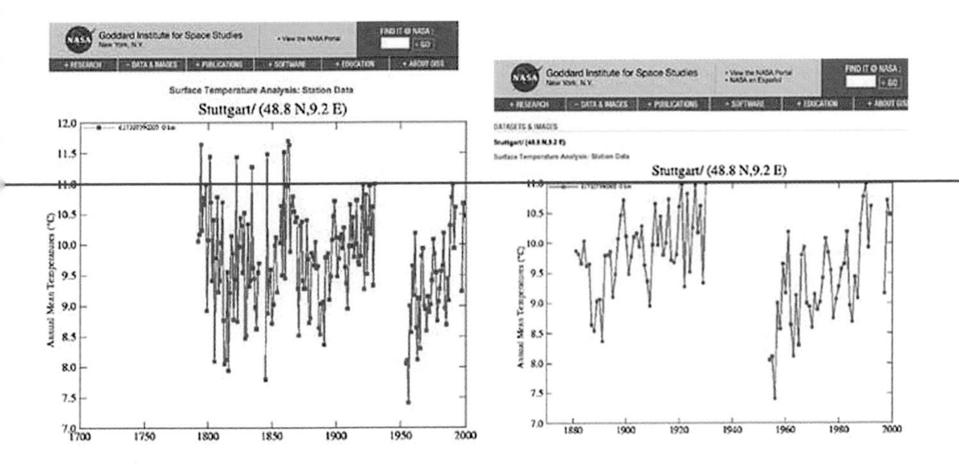

Fig. 7
Goddard Institute for Space Studies website, 2004 *and* 2007

But let's turn to the version on the right, which appeared when the website was redesigned in 2005, replacing the earlier example. Suddenly, both axes have been truncated, so the graphic shows a shorter time-frame, but more crucially, a narrower mean temperature span. The y-axis, which previously went up to 12 degrees C (in the version on the left), now only goes up to 11 degrees C, thus altering the story told by the data. Now, it appears, recent local warming in Stuttgart is just as fast and steep as at any other time in the historic past, the earlier spikes having been misleadingly cut off at a lower temperature than before.

Of course, all visual representations involve complex decisions about how to frame and manipulate the material, but this kind of deliberate refashioning of old data seems designed as an act of deception rather than clarification. And though most scientists and policy makers are rightly convinced by the mounting evidence for anthropogenic warming, it is still vital that this evidence is presented as honestly and objectively as possible — which is to say as *clearly* as possible — and where there is uncertainty, or missing data, or data that contradicts the wider picture, scientists should not be afraid of saying so.

The situation is severe enough, without having to massage or over-design the evidence, especially since this plays into the hands of oppositional interest groups who can point to such episodes and say (with some justification): 'look, the scientists are tampering with their findings'.

Such concern about the politicization of the scientific process goes back a surprisingly long way. When the British Association for the Advancement of Science was founded in 1831, its members were unanimous in declaring that statistics had no place in their organization, because the natural sciences, in their view, were theoretically based, value free, and impervious to political controversy, while the statistical sciences were none of those things, being principally governed by the political values of their practitioners. The first BAAS president, Adam Sedgwick, warned the statisticians that 'if they went into provinces not belonging to them, and opened a door of communication with the dreary world of politics, that instant would the foul demon of discord find his way into their Eden of philosophy.' Looking at the situation today, Sedgwick might well conclude that his forebodings were entirely justified.[10]

[10] Cited in Mary Poovey, 'Figures of Arithmetic, Figures of Speech: The Discourse of Statistics in the 1830s', in Chandler *et al* (eds), *Questions of Evidence* (1994), 401-21.

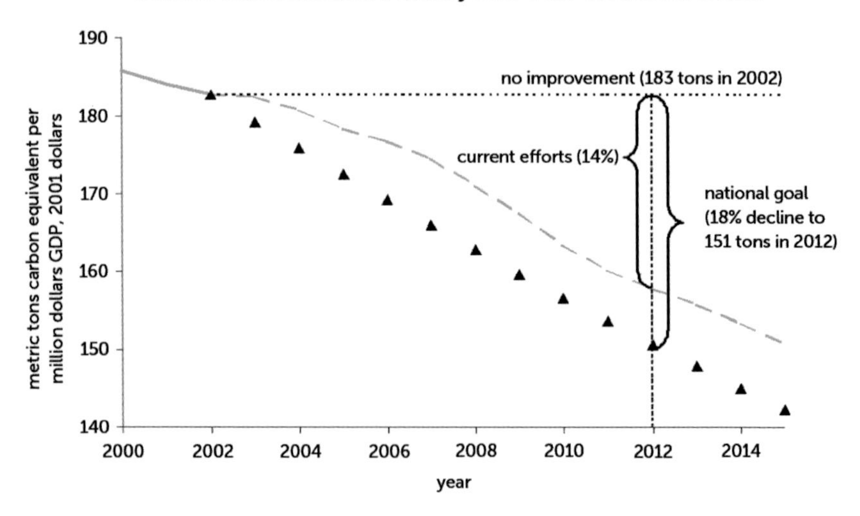

A particularly good example of the political misuse of graphically-presented data is the invention of the concept of Greenhouse Gas Intensity (GGI) by the United States administration in 2001, shortly after the election of President George W. Bush (see fig. 8).

Greenhouse Gas Intensity was, as the U.S. Environmental Protection Agency's website continues to define it, 'the ratio of greenhouse gas emissions to economic output'; in other words, it was not a quantity that could be directly measured. It was, instead, a ratio that related emissions to economic output. If, for example, a company one year produced 100 tonnes of carbon and 100 dollars' worth of goods, its greenhouse gas intensity would be one tonne per dollar. If the following year the same company produced the same amount of carbon, but an extra dollar's worth of goods, its GGI would have fallen by one percent. Even if it doubled its carbon emissions, a company — or a country — could claim a reduced intensity level as long as it more than doubled its output of goods.

Its supporters claimed that the Greenhouse Gas Intensity index was preferable to measuring straightforward emissions, because it recognized that a nation that continued to grow its economy was a nation that could afford to invest in sustainable technology. But the real value of GGI to its adherents lay in its visual pay-off: it turned the upward-trending graphic situation on its head, reversing the climate graph template established by the Keeling curve. Here, at last, was a *downward-trending* emissions scenario that could be brandished by the leaders of wealthy countries even as they continued to increase their emissions. The GGI register therefore gives a particularly sunny account of the U.S. situation, since between 1990 and 2000, U.S. Greenhouse Gas Intensity *fell* by seventeen percent, despite the fact that emissions actually *rose* by twelve percent. If emissions alone are measured, the United States is clearly among the world's

Fig. 8
Greenhouse Gas Intensity, from *U.S. Global Climate Change Policy: A New Approach* (Washington, D.C., February 2002)

worst offenders; but when you express the same data as GGI, it is immediately transformed into the most virtuous nation on earth, out-greening even the smallest and least carbon-hungry countries on the planet.

Greenhouse Gas Intensity has been described as a charade, and that is precisely what it is: a visual charade, with an alluring graphical story to tell — a downward-trending emissions graph —, but one that is entirely disingenuous. It writes an implied narrative of environmental improvement in the face of an actual narrative of continued despoilation. It is a visually consoling lie, designed to counteract the relentlessly upward-trending nature of climate science graphics. It is easy to imagine the sales pitch: the client (in this case, the United States government) demanded a downward-trending graphic, so: *voila*!, one duly appeared, cooked up from manipulated data in order to satisfy a visually simplistic and politically contaminated version of measurable atmospheric events.

* An article published in the *Philosophical Transactions* for October 1699 offered an attempt to formulate a mathematical model for assessing the personal credibility of eyewitness accounts. In a bewildering piece of applied mathematical logic, the anonymous author proceeded to calculate the confidence that could be invested in any first-person testimony:

> The Credibility of any Reporter is to be rated (1) by his Integrity, or Fidelity; and (2) by his Ability: and a double Ability is to be considered; both that of Apprehending, what is deliver'd; and also of Retaining it afterwards, till it be transmitted.

From this basis, the author goes on to factor in the differences in reliability between single and multiple eyewitness accounts, as well as between oral and written testimonies, and after some impressive calculations, he arrives at an average ratio of 5/6ths believability per person:

> Let there be Six Particulars of a Narrative equally Remarkable: If he to whom the Report is given, has 5/6ths of Certainty for the whole, or Summ, of them; he had 35 to one, against the Failure in any One certain Particular.
>
> For he has Five to One, there will be no Failure at all: And if there be, he has yet another Five to One, that it falls not upon that single Particular of the Six. That is, he has 5/6ths of Certainty for the whole: and of the 1/6th wanting he has likewise 5/6ths, or 5/36ths of the whole more; and therefore that there will be no Failure in that single Particular, he has 5/6ths and 5/36ths of Certainty, or 35/36ths of it.
>
> If Two Concurrent Reporters have, each of them, as 5/6ths of Certainty; they will both give me an Assurance of 35/36ths, or of 35 to one: if Three; an Assurance of 215/216, or of 215 to one,

and so on. Even if each of these eyewitnesses' stories was only 50 percent reliable, by this calculation, according to the author, two such witnesses would give you 3/4ths accuracy, three would give you 7/8ths, while ten would take you up to a level of 1023/1024ths accuracy. ('A Calculation of the Credibility of Human Testimony', *Philosophical Transactions* 21 (1699), 359-65.)

Because data can be so easily manipulated, the issue of what Theodore Porter called 'trust in numbers' remains problematic, while the personal credibility of whoever happens to be advancing an argument becomes ever more important; more important even than the supporting evidence.* 'The identification of trustworthy agents is necessary to the constitution of any body of knowledge', as Steven Shapin observed in *A Social History of Truth* (1994), but there is a particular atmosphere of mutual mistrust that surrounds the climate change debate, most notably in the United States, where anyone with something to say on the matter is assumed to have a vested interest. Even well-respected academics such as William Ruddiman, emeritus Professor of Environmental Sciences at the University of Virginia, feel it necessary to issue disclaimers such as the following, with which he ends his provocative history of climatic variation, *Plows, Plagues, and Petroleum*:

> I have never received any funding from either environmental or industry sources. All of my career funding has been from the government, and over 99 percent of it from the National Science Foundation, which is widely regarded by politicians of many views as the model of a well-run government funding agency (based on its reliance on competition and peer review). All of the funds used to write this book came from my retirement annuity earned at educational institutions.[11]

[11] William F. Ruddiman, *Plows, Plagues, and Petroleum: How Humans took Control of Climate* (Princeton, NJ: Princeton University Press, 2005), 177.

Ruddiman's declaration stems from the wariness of a researcher whose scientific objectivity has been called into question by those who do not care for his conclusions. 'The debate has taken a suprisingly ugly turn', he has said, evidently speaking from experience.

Ruddiman's emphasis on the peer-reviewed nature of his research funding is not surprising, given that peer review (the collegiate system by which new research is anonymously assessed and evaluated) has come to enjoy a near-hallowed status within mainstream science, where it is the principal conduit to publication in the top professional journals. But the cultural dominance of peer review has come under pressure in recent years, with critics of the system complaining that it is slow, elitist and inherently conservative, serving to suppress dissent against mainstream theories, while its supporters defend its consistent track record of sifting out methodological errors and faulty conclusions. Despite all this, as Professor Drummond Rennie (deputy editor of the *Journal of the American Medical Association* and an organizer of the International Congress on Peer Review and Biomedical Publication) has observed, even with peer review securely in place, 'there still seems to be no study too fragmented, no hypothesis too trivial, no literature too biased or too egotistical, no design too warped, no methodology too bungled, no presentation of results too inaccurate, too obscure, and too contradictory, no analysis too self-serving, no argument too circular, no conclusions too trifling or too unjustified, and no grammar and syntax too offensive for a paper to end up in print.'[12]

[12] Cited in the Fall 1991 issue of *ScienceWriters: The Newsletter of the National Association of Science Writers.*

Yet even the most conservatively peer-reviewed consensus position can still remain riddled with uncertainties. The mainstream consensus view on climate change, for example, as represented by the sequence of Assessment Reports published by the Intergovernmental Panel on Climate Change, is notable for its deployment of unusually wide uncertainty ranges, such as over the likely increases in average global temperatures in the event of carbon dioxide concentrations reaching twice their pre-industrial levels. These increases have been estimated as falling between 1.1 and 6.4 degrees C by the end of the century (according to the Fourth Assessment Report of 2007), with newspapers and environmental groups tending to cite figures from the top of the range, and oppositional lobby groups preferring to take figures from the bottom. Since neither extreme is as likely an outcome as a rise from somewhere within the middle range, there have been calls for more sophisticated uncertainty expressions to be used.

Stephen Schneider and Richard Moss, climate scientists who have referred to themselves as 'the uncertainty cops', have long argued that probabilities (such as 'a 60-70 percent likelihood') should be assigned to particular climate outcomes rather than the unhelpfully wide uncertainty ranges currently in use.[13] It was they who urged the IPCC to adopt the calibrated eight-point 'confidence scale' that appeared in the opening summary of the Third Assessment Report of 2001, and again in the Fourth Assessment Report of 2007 (see the chapter motto on page 45). They were less successful in arguing for the inclusion of their graphical representation of how varying degrees of confidence are arrived at. It took the form of a four-point axis, the compass points of which corresponded to confidence in: the theory; the observations; the models; and the consensus within the field. Once the coordinates are plotted they are joined up to make a distinctive shape, the area and outline of which serve to indicate the overall degree of confidence in the particular projected outcome, as well as revealing exactly *how* that degree of confidence has arisen (see fig. 9):

[13] Jim Giles, 'Scientific uncertainty: When doubt is a sure thing', *Nature* 418 (2002), 476-78.

Fig. 9
Schneider and Moss's graphical expression of confidence ranges in climate scenarios

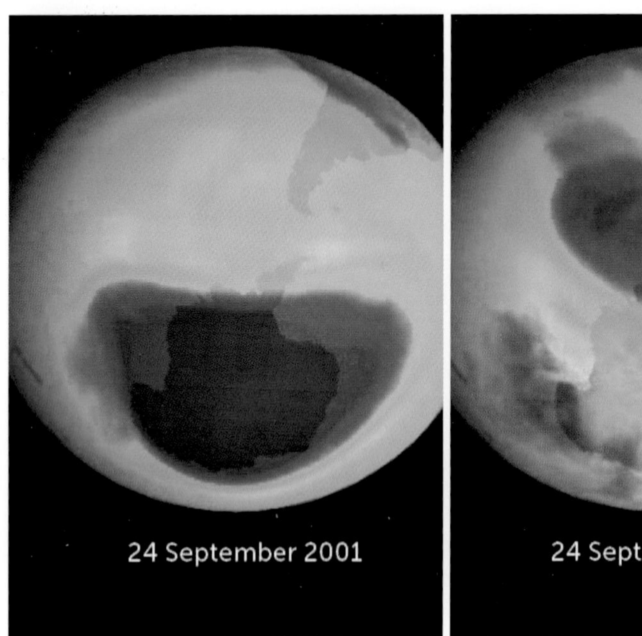

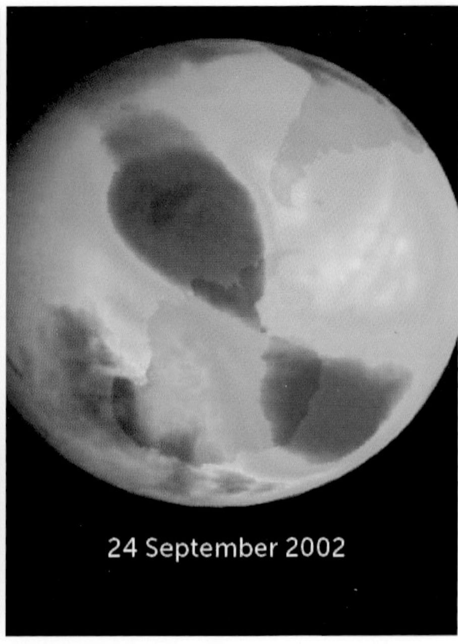

Fig. 10
The 'ozone hole' over Antarctica, as realized by the Goddard Space Flight Center's Scientific Visualization Studio. These images are among the most effective pieces of scientific visualization ever made.

As Theodore Porter argued in the opening pages of *Trust in Numbers*, 'let us suppose for the sake of argument that scientific investigation is able to yield true knowledge about objects and processes in the world. It must nonetheless do so through social processes. There is no other way. . . the norms of scientific communication presuppose that nature does not speak unambiguously.'[14]

[14] Theodore M. Porter, *Trust in Numbers: The Pursuit of Objectivity in Science and Public Life* (Princeton, NJ: Princeton University Press, 1995), 2.

And nature does not speak unambiguously — after all, we cannot actually see climate change; we can see some of its effects, of course, but not its ultimate causes. Increases in atmospheric greenhouse gases occur invisibly, and though they can be measured and plotted on scary-looking graphs, the fact of their invisibility remains a consistent barrier to action. If the sky changed colour as a result of increased carbon dioxide, there wouldn't be a problem now: emissions would have been halted long ago. But since the sky looks more-or-less the same no matter what we pump into it, what is needed is some kind of convincing visual representation of atmospheric change, such as the brilliantly coloured graphics that showed the ever-widening 'hole' in the ozone layer above Antarctica. First published in 1985 by NASA's Scientific Visualization Studio at the Goddard Space Flight Center, these false-colour graphics succeeded in visualizing an otherwise invisible process, and in such a striking and memorable manner that they served to alert the world to a looming environmental threat.

The process was remarkably quick: in the 1970s researchers at the British Antarctic Survey discovered that ozone concentrations above Antarctica appeared to be depleting, and as a consequence the level of ultraviolet (UV) radiation reaching the poles was steadily increasing. The culprit was soon identified: chlorofluorocarbons (CFCs), man-made compounds invented by industrial chemists in the late 1920s, and put to use in an array of applications, including refrigerators, air conditioners, solvents and propellants. By 1975 spray cans alone were pumping half a billion tonnes of them into the atmosphere, and by 1985, worldwide production of CFCs stood at almost two billion tonnes per year. But the problem with these compounds is that once they have been evaporated into the stratosphere, they are broken down by UV radiation, and their chlorine atoms released. Since chlorine can remain aloft for many decades, during which time a single atom can catalytically convert up to 100,000 ozone molecules into oxygen, even over the earth's mid-latitudes, it soon became apparent that the 'ozone hole' over Antarctica was only the opening rend in our planet's UV-protecting veil.

The wake-up call was a paper written by researchers at the British Antarctic Survey (and published in *Nature* in May 1985), in which the severity of the problem was demonstrated beyond doubt; NASA's false-colour visualizations soon began to appear, and the combination of rock-solid instrumental evidence and compelling graphic imagery galvanized the world into immediate action: the United States took the lead by banning the use of CFCs in spray cans, and a couple of years later, the United Nations Environment Program proposed the creation of a global convention guaranteeing the protection of the ozone layer. In 1987, this UN convention led to the signing of the Montreal Protocol: a treaty that deserves to be more famous than it is, since it represents humanity's first major victory over a global environmental crisis. Compared with the embarrassing political quagmire that was the Kyoto Protocol negotiations, the Montreal agreement was a model of effective international action, with every developed nation in the world willingly committing itself to the total elimination of CFCs from all industrial processes by 1996. A decade and a half on from that first Montreal deadline, and the ozone hole is already beginning to heal. It's a slow process, due mostly to the longevity of ozone-depleting chlorine, but at current rates ozone concentrations over the poles will have been restored to something like their pre-industrial levels by the middle of the present century.

No, nature does not speak unambiguously, and neither does instrumental data, and because of this environmental science will continue to require the kind of spectacular visualizations that led to the drafting of the Montreal Protocol. But climate scientists have so far failed to find an 'ozone hole' equivalent for global warming, and what we have ended up with instead is a mess of competing visual narratives characterized by suggestive shapes drawn by the plotted lines of story-laden graphs — the curves of which are easily controllable by artful restrictions on the kinds of information disclosed along their axes. 'And this is how the world ends: not with a bang, nor with a whimper, but with a PowerPoint presentation', as the *Guardian*'s Oliver Burkeman wrote at the Paris launch of the IPCC's Fourth Assessment Report in early February 2007, where a series of big, scary graphs, 'projected on to a vast screen above the heads of the assembled scientists . . . showed how global temperatures have skyrocketed in recent decades, and how they would skyrocket further in the immediate future: and they brought the words "hell" and "handcart" to mind.'[15]

[15] Oliver Burkeman, 'The scientists spoke cautiously but the graphs said it all', *Guardian*, 3 February 2007, 1.

The following terms have been used to indicate the assessed likelihood, using expert judgement, of an outcome or a result: Virtually certain >99% probability of occurrence; Extremely likely >95%; Very likely >90%; Likely >66%; More likely than not >50%; Unlikely <33%; Very unlikely <10%; Extremely unlikely <5% probability of occurrence.

IPCC Fourth Assessment Report, 2007

Climate change is the first major environmental crisis in which the experts appear more alarmed than the public. Most other environmental risk stories, from pesticides and 'global cooling' in the mid-twentieth century, to mobile phone masts and GM technology in the early twenty-first, have seen expert voices recruited to defuse the public's growing arsenal of technological fears. But when it comes to climate change, the response so far has been muted unease in the face of the escalating warnings of the scientists. As the ecologist Bill McKibben has observed, it may be that climate scientists have simply succeeded in adding 'another line to the long list of human problems — people think about 'global warming' in the same way they think about 'violence on television' or 'growing trade deficits', as a marginal concern to them, if a concern at all.'[16] In spite of their inescapable presence in the news schedules, climate change narratives must still compete in the overcrowded attention economy.

[16] Bill McKibben, 'Worried? Us?', *Granta* 83 (2003), 8.

NARRATIVES OF CLIMATE CHANGE

As was suggested in the previous chapter, one of the principal barriers to communicating climate change is its underlying invisibility, which has served to demote the growing crisis to a mere abstration, in spite of the profusion of visual 'evidence' to which our attention is constantly being drawn: melting ice-caps, shrinking rivers, encroaching deserts, rising waters. High visibility environmental crises such as hurricanes and floods — mostly isolated weather events that may or may not have any connection to wider climate change — have come to stand as emotive symbols of looming worldwide catastrophe.

Environmental coverage has always been characterized by a strong visual component, with events such as floods and oil spills being inherently more visually arresting than a notion as abstract as climate change. Such imagery, however, is often alarmist, and can lead to a rhetorical mismatch of scale between grim apocalyptic warnings and the cheerfully mundane solutions proffered as useful mitigating actions: 'use less hot water'; 'check your tires'; 'plant a tree', as Al Gore suggests in his list of '10 Things to Do' which appears at the end of *An Inconvenient Truth*, in marked contrast to the end-of-the-world rhetoric that informs the rest of the film. Such bland assurances can actually make the problem seem more, rather than less, overwhelming, for how on earth will half-filling our kettles have any impact on a climate regime that is already committed to several degrees of future warming, regardless of any measures we might put in place today?

A report published in August 2006 by the Institute for Public Policy Research, entitled *Warm Words: How are we telling the climate change story and can we tell it better?*, criticized the outright alarmism of much media coverage of climate change, arguing that the employment of a quasi-religious register of doom served to exclude the possibility of any meaningful action on the part of the reader or viewer. The more we are bombarded with predictions of the dire effects of runaway warming, the more likely we are to switch off altogether, retreating into helplessness, outright scepticism, or something that the authors of the report called 'British comic nihilism… a sunny refusal to engage in the debate.'[17]

[17] Gill Ereaut and Nat Segnit, *Warm Words: How are we telling the climate change story and can we tell it better?* (London: IPPR, 2006), 15. A follow-up report, entitled *Warm Words II: How the climate story is evolving and the lessons we can learn for encouraging public action*, appeared in September 2007.

In the United States, by contrast, climate change is often perceived as a global engineering problem in search of top-down technological solutions, while in northern Europe, particularly in Britain, it is often cast as a moral question, to which every individual is urged to contribute a share of responsible action. These historic attitudes have given rise to something of a mitigation culture clash, in which large-scale geo-engineering solutions, such as cloud seeding, ocean fertilization, or the switch to a hydrogen fuel economy, are ranged against the more abstract virtues of reducing personal energy consumption — often expressed in the form of the so-called 'carbon footprint', a complex spatial metaphor that translates notional volumes of emitted carbon dioxide and other invisible greenhouse gases into easily comparable surface areas (fig. 11).

Fig. 11
Milk's carbon footprint: a complex spatial metaphor expressing the CO_2 (by weight) released during the production process. Note quaint mix of metric and imperial units.

But while the proponents of these competing approaches continue to offer divergent ideas, wide-ranging creative approaches continue to be in short supply. The reality of ongoing climate change has yet to be embraced as a stimulus to creativity — in the arts as well as the sciences — or as a permanent and inescapable part of human societal development. It is not simply a looming technical problem to be fixed, as Mike Hulme of the Tyndall Centre for Climate Change Research has put it:

> Climate change is not a problem waiting for a solution. Engineers are very useful people, but they are not going to give us the answers here... we need a far richer array of intellectual traditions and methods to help us analyse and understand the problem — behavioural psychologists, sociologists, faith leaders, technology analysts, artists, political scientists, to name a few.[18]

[18] Mike Hulme, 'MicroTate Environment Special', *TateEtc* 9 (2007), 109.

And though the success of the 1987 Montreal Protocol offers a compelling precedent for strongly visualized evidence leading to swift and effective international action, climate change presents a more technically challenging, historically layered, and rhetorically complex set of problems, many of which were already present at the very outset of the story, when, on 11 December 1895, the Swedish chemist Svante Arrhenius read a landmark paper at the Royal Swedish Academy of Sciences.

Arrhenius's now famous paper — 'On the Influence of Carbonic Acid in the Air upon the Temperature of the Ground' — outlined the likely impact that increased concentrations of carbon dioxide ('carbonic acid') in the atmosphere would have on the surface temperature of the earth. The paper, published in an English translation the following year, pointed out that the earth's heat budget was greatly influenced by the presence of trace constituents in the atmosphere, notably carbon dioxide, water vapour, ozone and hydrocarbons. Using the published results of John Tyndall's 1860s research into the heat-absorbing and radiative properties of atmospheric gases, Arrhenius had calculated (painstakingly, by hand, over the course of a year of fourteen-hour days in which he laboured to overcome the heartbreak of losing his wife to another man) that the removal of all atmospheric carbon dioxide would cause the earth's temperature to drop by at least 20-30 degrees Celsius, and, conversely, that the doubling of atmospheric CO_2 (from its 1890s concentrations of *c.* 300 parts per million) would cause average global temperatures to rise by around 5 degrees C, with the greatest increase being seen at the poles, where 'the temperature of the Arctic regions would rise about 8 degrees or 9 degrees Celsius.'[19]

[19] Svante Arrhenius, 'On the Influence of Carbonic Acid in the Air upon the Temperature of the Ground', *The London, Edinburgh and Dublin Philosophical Magazine and Journal of Science* 41 (1896), 237-76.

Arrhenius had first embarked on these calculations as part of his research into the cyclical nature of the earth's ice ages, but their implications had soon led him towards the novel prediction that significant increases in industrial emissions would contribute to future rises in average global temperatures. As he went on to argue in *Worlds in the Making*, a widely translated work of popular science that first appeared in English in 1908, 'the percentage of carbonic acid [CO_2] in the air must be increasing at a constant rate as long as the consumption of coal, petroleum, etc., is maintained at its present figure, and at a still more rapid rate if this consumption should continue to increase as it does now.'[20] Arrhenius went on to claim that the increased burning of fossil fuels could even lead to temperature rises high enough to avert the earth's next ice age:

[20] Svante Arrhenius, *Worlds in the Making: The Evolution of the Universe*, trans. H. Borns (London & New York: Harpers, 1908), 57-58.

> Is it probable that we shall in the coming geological ages be visited by a new ice period that will drive us from our temperate countries into the hotter climates of Africa? There does not appear to be much ground for such an apprehension. The enormous combustion of coal by our industrial establishments suffices to increase the percentage of carbon dioxide in the air to a perceptible degree.

Arrhenius also responded to what he described as 'lamentations that the coal stored up in the earth is wasted by the present generation without any thought for the future', by suggesting that the likely doubling of atmospheric CO_2 over the next two to three thousand years could also lead to a period of global runaway warming that would allow the world's population 'to enjoy ages with more equable and better climates, especially as regards the colder regions of the earth, ages when the earth will bring forth much more abundant crops than at present, for the benefit of rapidly propagating mankind.'[21]

[21] Ibid., 61-63.

What is notable here is that Arrhenius's description of a hypothetically warming world is already preoccupied by many of the issues around which climate change is debated today. These include the consequences of increased reliance on finite fuel resources; the stresses of global population increase; the mass migration of climate refugees; the prospect of human actions exerting long-term influence over weather and climate; and the concept of major feedback mechanisms in nature. All have become central conceits within the twenty-first-century debate, yet all seem to have emerged fully formed at the moment when the novel hypothesis of anthropogenic warming was first being tentatively proposed. What renders this so striking, at least from a science history perspective, is that Arrhenius's predictions were directed towards a hypothetical distant future, and had not been made in reaction to any measurable consequences evident in the nineteenth-century present. He arrived at his conjectures by imagining the effects of long-term chemical processes in the atmosphere, rather than through any kind of direct observation.

In other words, Arrhenius's hypotheses were shaped by conjectural imagery, his projections constructed and argued through what amounted to imaginative literary conceits. Such conceits were characteristic features of nineteenth-century scientific popularization, yet Arrhenius's hypotheses have continued to shape the terms of the present debate. As Sir Nicholas Stern pointed out in his economic *Review*, the parameters of Arrhenius's calculations may have changed over the past one hundred years, but the central, hypothetical image of human-enhanced runaway warming that was first introduced in 1895-6 remains powerfully influential, even if 'the atmosphere is much more complicated than [his] simple models suggest.'[22]

[22] Nicholas Stern, *The Economics of Climate Change: The Stern Review* (Cambridge: CUP, 2007), 7.

Perhaps it's inevitable that Svante Arrhenius, the progenitor of so much climate change imagery, has himself become an environmentalist icon, fêted as 'the father of climate change science', and the founder of a century-old rhetorical lineage that extends into the present. In spite of his lesser-known conclusions concerning the future benefits of an all-out anthropogenic warming phase, the iconography of Arrhenius as the father of cautionary climate science has served to place him at the head of a line of climate change whistleblowers, his hypothetical conjectures repurposed as a prescient warning to mankind.

The figure of the whistleblower — the concerned individual who takes on the vested interests of a corporate consensus — has become something of a fixture in environmental communication, particularly in contested policy areas, such as nuclear power, pesticides, and climate change.[23] Bill McKibben has described how climate scientists and activists inhabit 'one of those strange dreams where the dreamer desperately needs to warn someone about something bad and imminent; but somehow, no matter how hard he shouts, the other person in the dream — standing smiling, perhaps, with his back to an oncoming train — can't hear him.'[24] For a long time scientists maintained that they have no business making political statements, but climate change has removed such historic inhibitions, and now the course of the debate is led by instances of direct intervention. James E. Hansen's appearance before the U.S. Senate Committee on Energy and Natural Resources on 23 June 1988 was the first and most high profile such episode, and represents an environmental communications watershed. Hansen, director of NASA's Goddard Institute of Space Studies, told the Senate that 'global warming has begun', claiming that he was '99 percent confident' that rising temperatures represented a warming trend rather than any kind of natural variability.[25] Hansen's graphic warning that 'we are loading the climate dice' received unprecedented levels of attention from the media and from policy makers partly because it coincided with a record-breaking heatwave in the United States. As environmental sociologists have often noted, a single visible event or phenomenon often stands in for something much wider, and the stifling heatwave of 1988 supplied an ideal context for the launch of 'global warming' as a public issue in the States.[26]

[23] The figure of the 'concerned scientist' is discussed in Robert Cox, *Environmental Communication and the Public Sphere*, 2nd edn (Thousand Oaks, Ca.: Sage Publications, 2009), 351-56.

[24] McKibben, 'Worried? Us?', 8.

[25] For the full text of Hansen's deposition see *The Hearing of the Committee on Energy and Natural Resources, U.S. Senate, 100th Congress, June 23, 1988* (Washington D.C.: U.S. G.P.O., 1988), II, 39-41.

[26] See J. Shanahan and J. Good, 'Heat and hot air: influence of local temperature on journalists' coverage of global warming', *Public Understanding of Science* 9:3 (2000), 285-95.

As an indirect consequence of Hansen's testimony, the image of the lone voice has come to occupy a central position at the heart of the global warming story. Historical accounts of the subject tend to hinge on moments of individual testimony, the 'lone voice in the greenhouse', as a headline in *Nature* dubbed the early twentieth-century climate scientist Guy Stewart Callendar.[27] Along with Hansen and Arrhenius, Callendar tends to be invoked early on in historical accounts of climate change, Spencer Weart's *The Discovery of Global Warming*, for example, beginning with an admiring description of Callendar's 1930s research into the phenomenon of human-enhanced warming. Unlike that of Arrhenius, Callendar's work was undertaken in response to an observed twentieth-century warming phase, although in common with Arrhenius, Callendar was on his own when it came to ascribing a central role to the increased combustion of fossil fuels. Weart's account of a lecture delivered by Callendar to the Royal Meteorological Society in 1938 drew less on the precedent of Arrhenius's earlier paper than on a series of associated ideas concerning the integrity of the unbiased amateur. 'One man challenged the consensus of the experts', he wrote:

[27] Robert J. Charlson, 'A lone voice in the greenhouse', *Nature* 448 (19 July 2007), 254.

> Callendar was out of place, for he was no professional meteorologist, not even a scientist, but an engineer who worked on steam power. He had an amateur interest in climate and had spent many hours of spare time putting together weather statistics as a hobby. He had confirmed (more thoroughly than anyone else) that the numbers indeed showed global warming. Now Callendar told the meteorologists he knew what was responsible. It was us, human industry. Everywhere we burned fossil fuels we emitted millions of tons of carbon dioxide gas (CO_2), and that was changing the climate.[28]

[28] Spencer Weart, The *Discovery of Global Warming*, 2nd edn (Cambridge, Mass.: Harvard University Press, 2008), 2; see also James R. Fleming, *The Callendar Effect: The Life and Work of Guy Stewart Callendar (1898-1964), the Scientist who Established the Carbon Dioxide Theory of Climate Change* (Boston, Mass.: American Meteorological Society, 2007).

Weart's account makes direct appeal to the image of the doughty British engineer as a man whose professional successes derive from defying 'the consensus of the experts', conformists whose only role in the story is to claim that something can't be said or done. The scope of Callendar's labours was impressive given his status as an amateur meteorologist (he was by profession a steam engineer attached to the British Electrical Research Association). Another of life's lonely statistical enthusiasts, Callendar compiled a vast body of global temperature data, including those taken from his own Sussex garden, and applied it to measurements of rising concentrations of atmospheric carbon dioxide since pre-industrial times, as well as to recent and historical accounts of observed glacial retreat. Callendar's great insight, that recent glacial loss might be connected to increased fossil fuel use, was by far the most influential idea that he introduced — the icon of melting ice having proved the most enduring single climate change image of the past thirty years — though like Arrhenius before him, Callendar was unperturbed by the prospect, and ended his paper by expressing the hope that 'the return of the deadly glaciers should be delayed indefinitely' by runaway global warming.[29]

[29] Guy S. Callendar, 'The Artificial Production of Carbon Dioxide and its Influence on Climate', *Quarterly Journal of the Royal Meteorological Society* 64 (1938), 223-40.

Although the last few decades have seen the rise of a near-universal consensus concerning the reality of the threat posed by global climate change, the image of the lone voice remains central to the story, even if the role has changed to one of endorsing rather than defying the consensus view. This is especially the case now that policy makers have started to determine the direction of the climate change debate. In an article published in *Science* in January 2004, for example, Professor Sir David King (at the time the UK government's Chief Scientific Adviser) famously described climate change as 'the most severe problem we are facing today — more serious even than the threat of terrorism', his comments quickly finding their way into newspaper headlines across the world.[30] By setting up a direct rhetorical contest between the two leading narratives of the age, King's soundbite had gifted news editors with a scientifically accredited promise of something more terrifying than terror, and the comparison was quickly established as a fixture of climate change discourse. 'Is it possible that we should prepare against other threats besides terrorists?', as Al Gore asked in the cinema trailer to *An Inconvenient Truth* (2006), the question accompanying a computer generated image of the World Trade Center Memorial Site being engulfed by a rising sea. As one icon of global terror was slowly erased by another, the policy implication was plain

[30] David A. King, 'Climate Change Science: Adapt, Mitigate, or Ignore?', *Science* 303 (2004), 176-77.

to see: where is the corresponding global war on warming? — a question that has been asked many times since, most directly by the astrophysicist Stephen Hawking, who said: 'Terror only kills hundreds or thousands of people. Global warming could kill millions. We should have a war on global warming rather than the war on terror.'[31]

[31] 'US must win the war on climate change, says Charles', *Guardian*, 29 January 2007; Hawking quoted in the *Times*, 31 January 2007, 3.

What happened a few weeks after the comparison was first published, however, overshadowed any kind of policy debate that King might have hoped to initiate. At the American Association for the Advancement of Science conference, held in Seattle in February 2004, a freelance science reporter named Michael Martin happened to pick up a computer disk that had been left behind at the end of a roundtable discussion at which King had been a participant. When he opened the disk Martin found a copy of a memo from Ivan Rogers (Prime Minister Tony Blair's chief private secretary), which politely but firmly advised King not to make any further statements comparing global warming with global terrorism.[32] The memo then went on to give detailed instructions on what to say should King find himself being questioned on the matter by American reporters, including scripted answers to 136 anticipated questions (Downing Street researchers work fast). If asked, for example, to compare estimated numbers of future deaths likely to be caused either by climate change or terrorism, King was instructed to say: 'the value of any comparison would be highly questionable — we are talking about threats that are intrinsically different', and if pressed on which of the two he considered the greater risk to life, his answer was to be the noncommittal: 'both are serious and immediate problems for the world today.' King's terrorism comparison was, in effect, to be edited out of the climate change repertoire, since, in Downing Street's words, it 'distracts from our wider efforts to engage the U.S. on climate change... this kind of discussion does not help us achieve our wider policy aims.'

[32] Michael Martin, 'Cooler Heads on Climate Change', *ScienceNow* 217 (2004), 2 (sciencenow.sciencemag.org/cgi/content/full/2004/217/2).

The story of the accidentally leaked directive was given prominent coverage in the British press, much of which recycled the image of the lone whistleblower, casting King in the role of the dedicated expert who had stumbled across something alarming, only to find himself abruptly silenced by his shadowy political paymasters — 'BLAIR SCIENCE ADVISER 'GAGGED' BY NO 10 AFTER WARNING OF GLOBAL WARMING THREAT' (The *Independent*); 'DOWNING ST 'GAGS CHIEF ADVISER ON GLOBAL WARMING'' (The *Daily Telegraph*) — in a manner reminiscent of the coverage of the death of the UN weapons inspector David Kelly a few months before. In both cases the press went out of its way to portray the individual scientist as a man of integrity, whose unbiased reading of 'the evidence' had forced a dramatic confrontation with the political machine, an image long familiar from classic whistle-blower narratives such as James Bridges's *The China Syndrome* (1979), Mike Nichols's *Silkwood* (1983) or Michael Mann's *The Insider* (1999). King has described the aftermath of the 'gagging' episode as 'exactly what we hoped for — multiplied by 100', his original dig at U.S. climate change policy having gone on to become a globally familiar soundbite as a result of the press furore, though he was (perhaps not surprisingly) never invited back to the White House.[34]

[34] Telephone interview with RH, 10 February 2009. In the same conversation King also disclosed that he had turned down the role that was subsequently taken by Al Gore in *An Inconvenient Truth*.

The iconography of the whistleblower continues to be a feature of climate change narratives, whether supportive of the consensus view or not. In fact, many leading climate change sceptics such as the Danish statistician Bjørn Lomborg, or the American physicist S. Fred Singer, have sought to appropriate and repurpose the role, portraying themselves as lone, sane voices pitted against the unthinking consensus promoted by self-interested scientists. William Ruddiman, in his *Plows, Plagues and Petroleum*, has described this reversal as 'the "white knight" or "hero" syndrome', the conviction that only heroic action in uncovering the truth will save humanity from disaster or folly:

> many contrarians appear to see mainstream scientists as dull-witted sheep following piles of federal grant money doled out by obliging federal program managers. In this view, only those who toe the party line that the global-warming problem is real, large, and threatening

will get their hands on federal money. And of course only the lone visionary with clear vision can save the day.[35]

[35] Ruddiman, *Plows, Plagues, and Petroleum*, 187.

Meanwhile, natural-world understudies for the role of whistleblower are regularly supplied by what could be characterized as climate change canaries: individualized instances of warning signs or wake-up calls, alerting the world to the presence of unseen perils. In the context of global climate change such canaries have usually been glaciers or icecaps, seen either in retreat or in dramatic fragmentation — as in the case of the Larsen B ice-shelf off the Antarctic Peninsula, a 720-billion-tonne apron of ice that broke up over the course of a month in early 2002 — or they are examples of a single threatened or displaced species, such as the starving polar bear filmed swimming through what ought to have been a solid platform of spring ice.

But as critics such as Bjørn Lomborg enjoy pointing out, while the sight of a hungry bear swimming through slush is highly emotive, polar bear numbers are in fact holding pretty steady: in 1960 there were approximately 5,000 arctic bears; now (in the wake of limits on hunting) there are around 20,000: far from being endangered, they are actually experiencing an age of bear glut, and hundreds of bears a year are shot, mainly in Alaska and Canada. However, as Jon Adams has suggested, Lomborg seems to have simply totted up polar bear numbers regardless of their geographical spread, and many of the bears shot each year are likely to be scavengers encroaching on towns far south of their natural hunting grounds, where the living is easy, and where, like urban foxes, their numbers will naturally increase. So the question for us is: is a growing polar bear population dependent on the contents of Alaskan landfills of greater or lesser ecological 'value' than a far smaller population of roaming bears acting as we prefer them: as top predators in a stable Arctic ecosystem?[36]

[36] Jon Adams, eyewitness account of Bjørn Lomborg's lecture at the LSE on 2 October 2007.

In their role as climate change canaries, clouds are both more prevalent and closer to home than Arctic ice-caps or polar bears. Cloud patterns have long been read as short-range weather indicators, but more recently they have begun to be seen as longer term climatic signals. Their messages are far from clear, however, and so far little is certain about the roles that clouds are likely to play in shaping future conditions on earth. Will clouds turn out to be agents of warming, veiling us in an ever-thickening blanket of emissions, or will they end up saving the day by reflecting ever more sunlight back into space? These, it turns out, are far from simple questions, and cloud behaviour continues to offer serious impediments to understanding future climates, since a change in almost any aspect of clouds, such as their type, location, water content, longevity, altitude, particle size and overall shape, changes the degree to which clouds will serve to warm or cool the earth.

As is so often the case with climate science, research yields apparently contradictory results. On the one hand, for example, many climate scientists believe that continued surface warming will cause increased water vapour to rise from the oceans, leading to an overall increase in cloud formation — while on the other hand, particularly in warmer latitudes, an increase in the water vapour content of our atmosphere could see large convective cumuliform clouds building up and raining out far quicker than they do at present, thereby leading to a net *decrease* in the earth's total cloud cover. Low-level stratiform clouds, meanwhile, tend to shield the earth from incoming solar radiation, but recent research has shown that such clouds are more likely to dissipate in warmer conditions, thus allowing the oceans to heat up further, and causing yet further stratiform cloud loss.[37] Scientists currently have no idea which of these outcomes is the most likely, nor do they really know the kind of long-term influences that either is likely to have. Even if, for the sake of argument, it's assumed that overall cloud cover will increase as the surface of

[37] Amy C. Clement *et al*, 'Observational and Model Evidence for Positive Low-Level Cloud Feedback', *Science* 325 (24 July 2009), 460-64.

our planet continues to warm, it remains unclear what kind of clouds (and thus what kind of feedback mechanisms) are likely to predominate.

For instance, high, thin clouds, such as cirrostratus, tend to have an overall warming effect, as they admit shortwave solar radiation in from above, while bouncing longwave back-radiation (reflected from the sunlit ground) back down to earth. Any increase in cirrostratus cloud cover would therefore add another warming mechanism to our climate. In contrast, however, bright, dense cumulus clouds serve to cool the earth by reflecting incoming sunlight back into space by day. At night, these same clouds tend to exert a slight warming effect, by absorbing or reflecting back-radiation, but their overall influence is a cooling one, especially when their summits grow dense and white. So, in theory, an increase in high, thin clouds would amplify the global warming effect, while an increase in low, dense, puffy clouds would have a contrary cooling effect — which is why cloud-whitening has recently been advanced as a geo-engineering idea for mitigating the effects of climate change, with salt water to be sprayed from thousands of ships in order to create brighter and more reflective clouds over the oceans. In reality, of course, things are never that simple, and clouds have always had an interesting habit of behaving in unpredictable ways.

For example, after the terrorist attacks of 11 September 2001, all commercial flights in the United States were grounded for several days, leaving the skies contrail-free for the first time in decades. The result, according to a comparison of nationwide temperature records, was slightly warmer days and slightly cooler nights than were usual for that time of year, the normal night/day temperature range having increased by 1.1 degrees C. According to the climate scientists who worked on the data, this was probably due to additional sunlight reaching the surface by day, and additional radiation escaping at night through the unusually cloudless skies.[38] At first sight this might seem counter-intuitive, for surely the

[38] David J. Travis *et al*, 'Contrails reduce daily temperature range', *Nature* 418 (2002), 601.

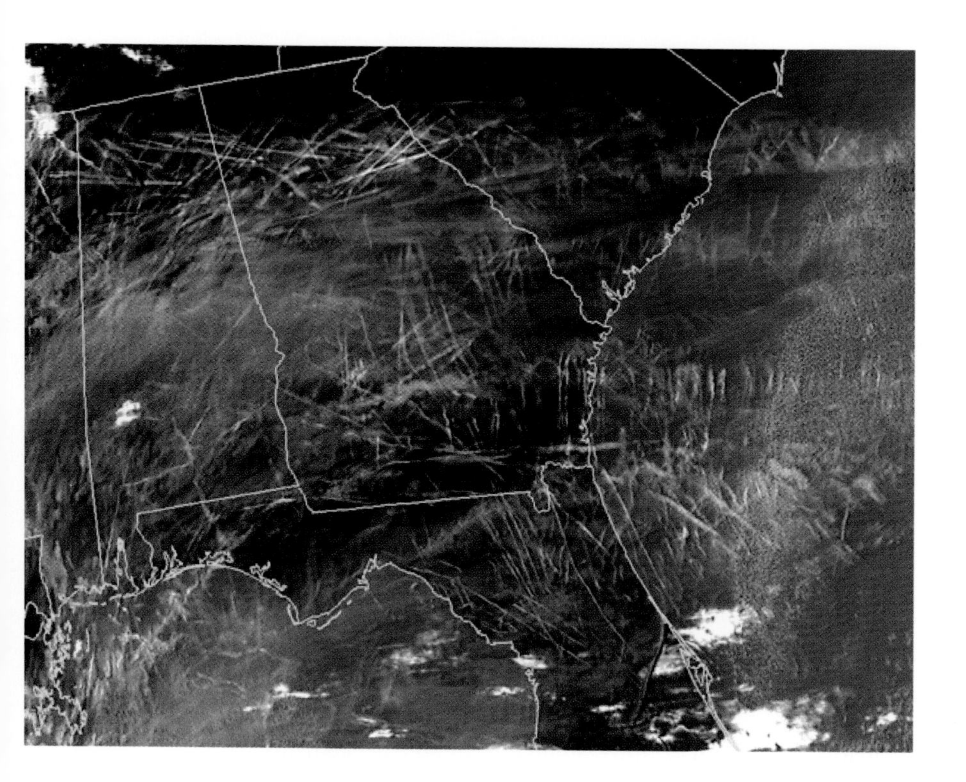

Fig. 12
Contrails over Southeastern USA, 29 January 2004 (NASA)

kind of cirriform clouds created by the spreading of aircraft contrails are straightforward warming clouds, the kind that allow sunlight through, while bouncing back-radiation down to the lower atmosphere? Surely an absence of contrails ought to have an overall *cooling* effect?

But contrails are a lot more complicated than that, because when they are in their initial, water droplet, stage they are denser than natural cirrus clouds, since they are created from two distinct sources of vapour: the moisture emitted by the aircraft's exhaust, and the moisture already in the atmosphere, all of which is condensed into a turbulent mixture of large water droplets and ice crystals, seeded on the solid particulates present in the exhaust plume. At first, this young contrail behaves more like a fluffy low level cloud, reflecting sunlight back into space, and exerting a short-term localized cooling effect. But if persistent contrails start to spread, they thin out into cirriform cloud layers, which can often cover large areas of sky. Their overall effect then reverts to a warming one, consistent with the known behaviour of natural cirriform clouds.

The picture is complicated yet further by the time of day that the contrails form and spread. If contrails spread during the early morning or late evening, they can exercise a slight cooling effect, due to the angle at which sunlight is reflected off the ice crystals into the upper atmosphere. At night, by contrast, all clouds, including contrails, can only exert a warming effect, since there is no incoming sunlight to reflect into space. Any increase in night flights is therefore likely to raise temperatures on the ground: and that increase is already well underway. In fact, the projected warming effects associated with the rise in night flights are in the region of a 0.2-0.3 degrees C hike per decade in the United States alone — and this figure does not include the other warming effects of aviation, such as increased CO_2 emissions and local ozone formation.[39] Of course, much about contrail science remains new and uncertain, and little about these man-made clouds is understood entirely, especially when it comes to the skies above the developing world, where flights are becoming increasingly prevalent. But the difference between the skies above busy flight corridors and those above sparsely flown areas is clearly visible from space (see fig. 12). Whether aircraft of the future will need to change the altitudes or times of day at which they fly in order to modify their contrail formation is a matter of current speculation; as David Travis, the atmospheric scientist who led the post-9/11 contrail research, has pointed out, 'what we've shown is that contrails are capable of affecting temperatures. Which direction, in terms of net heating or cooling, is still up in the air.'

[39] Patrick Minnis *et al*, 'Contrails, Cirrus Trends, and Climate', *Journal of Climate* 17:8 (2004), 1671-85

Equally up in the air, albeit at a far greater distance, are noctilucent clouds (NLCs), the changing patterns of which have become apparent over the past two decades (see fig. 13). First observed and named in the 1880s, NLCs were once the rarest

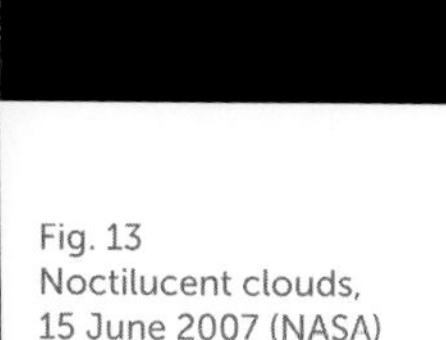

Fig. 13
Noctilucent clouds, 15 June 2007 (NASA)

clouds of all, but not only are they now appearing far more often, they also shine brighter than they did before, and are observable from increasingly lower latitudes.[40] According to one hypothesis, NLCs are being formed from plumes of space shuttle exhaust jettisoned into the earth's upper atmosphere, where neither water vapour nor dust nuclei are common natural occurrences, and therefore these clouds' increased appearance (an increase of 8 percent per decade) is due to a proportionate increase in space shuttle traffic. Other research, however, points to the fact that extreme cold is needed to form icy clouds in environments as dry as the mesosphere, 50 to 80 kilometres above the earth's surface, where temperatures as low as -130 degrees C are normal.

Strange as it may seem, the increased concentrations of atmospheric greenhouse gases that have contributed to raising temperatures on earth are also serving to create *colder* conditions in the earth's outer atmosphere. This is because greenhouse gases trap much of the longwave surface radiation that has started its return journey back out into space. With less thermal energy able to escape from the lower atmosphere,

[40] Sourish Basu, 'Clouds that Rival Auroras Now Bigger and Brighter', *Scientific American Online News*, 10 July 2007.

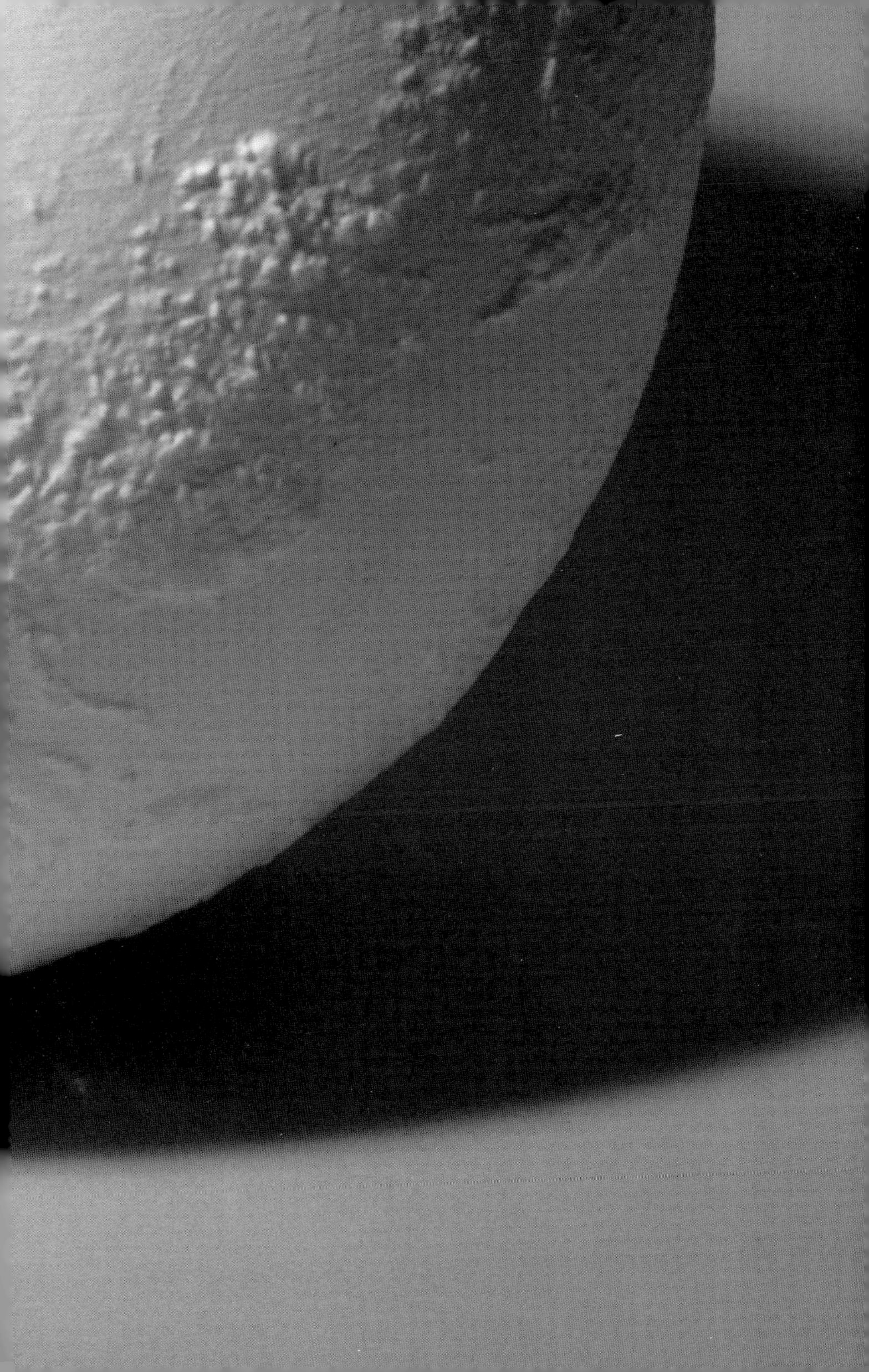

Fig. 14
Preceding page:
A Planetary Order (Terrestrial Cloud Globe), Martin John Callanan

the upper atmosphere is thereby growing correspondingly chillier. So could the observed increase in noctilucent cloud formation be due to mesospheric cooling, the lesser-known counterpart to global surface warming; and might their increased brightness be due to larger ice crystals being formed from a high-altitude influx of water vapour from the warming layers below? After all, NLCs have only been in evidence since the 1880s, the heyday of the Industrial Revolution, so it is possible that they will turn out to be yet another anthropogenic phenomenon — if so, the visible impact of human activity will have extended much further into our fragile atmosphere than we could ever have previously suspected. Whatever the secrets of these mysterious clouds, it is hoped that the AIM (Aeronomy of Ice in the Mesosphere) satellite, which was launched by NASA in April 2007 on a mission to study NLCs at close range, will be able to provide some answers to these questions.

Viewed from ground level, clouds are short-lived localized phenomena, undergoing rapid alterations as they pass overhead; when viewed from space, however, their individual movements are subsumed into large-scale formations that range slowly across the earth's surface, connecting vast tracts of land and sea through enormous geophysical processes. Seen from space, what from earth is merely an indistinct bank of stratocumulus cloud, becomes part of a visible planetary order. It was this dual perspective that led Martin John Callanan to produce a terrestrial cloud globe, entitled *A Planetary Order*, the many technical challenges of which were worked through and overcome during his residency at the UCL Environment Institute.

'Unlike Richard, who's got a huge fascination with clouds, I'm more interested in systems — systems that define how we live our lives' (MJC). Showing the earth's cloud cover from one second in time, the shimmering white cloud globe freeze-frames the entire operation of the global atmospheric regime, and highlights how fragile the environmental (and informational) systems are that operate across the world. For the globe is created from raw information, being a physical visualization of real-time scientific data. One second's worth of readings from all six cloud-monitoring satellites that are currently overseen by NASA and the European Space Agency was transformed into a virtual 3-D computer model, which was '3-D drawn', or rather, laser melted, at the Digital Manufacturing Centre at the UCL Bartlett Faculty of the Built Environment. It was the largest object ever created by the Digital Manufacturing Centre, and it took two full days to build, the delicate outlines and profiles of the clouds emerging slowly as the laser carved gently across the compacted nylon powder surface of the sphere.

Unlike most of NASA's own data visualizations, the globe features no added colour, only the sculpted whiteness of the raw material that throws a maze of faint shadows across the structure. From out of these shadows, in the right angles of light, emerge the global cloud patterns as captured on 2 February 2009 at 0600 UTC precisely, and, under them, the implied outlines of the continents below, seen as though glimpsed through mist, or rather, through the mystifying quantity of atmospheric data that is currently being collected from the silent fleet of satellites in orbit some 36,000 kilometres out in space — an increasingly hertzian environment, where an electronic Babel of satellites, radio signals, text messages and security frequencies vibrates, day and night, with invisible streams of man-made weather. Though far from earth's surface, we have nevertheless made it back to something resembling Borges's 1:1 scale Map of the Empire, for, by taking a single second's worth of transmitted information, our entire world has been made anew, pristine, white, and wreathed in the haze of an artificial atmosphere, held aloft like the fossilized egg of a long-extinct species that is about to be brought back to life from a single rescued strand of DNA.

Figs. 15-24 and 25, 26 on the following four pages
A Planetary Order (Terrestrial Cloud Globe), Martin John Callanan

IS QUADRANGLE OF UNIVERSITY COLLEGE LONDON
HICH THE FIRST STONE WAS LAID ON 30 APRIL 1827
BY H R H THE DUKE OF SUSSEX
DECLARED TO BE COMPLETE ON 15 NOVEMBER 1985
BY H M QUEEN ELIZABETH II

Information which is not held because it never existed is covered by exception 12(4) (a) and must be dealt with correctly under Regulation 14.

Defra, Environmental Information Regulations, 2004

TEXT TRENDS

Over the past twenty years, global climate change has emerged as the overarching narrative of our age, uniting a series of ongoing concerns about human relations with nature, the responsibilities of first world nations to those of the developing world, and the obligations of present to future generations. But if the climate change story entered the public realm as a data-driven scientific concept, it was quickly transformed into something that the ecologist William Cronon has called a 'secular prophecy', a grand narrative freighted with powerful, even transcendent languages and values. And though climate science can sometimes adopt the rhetoric of extreme quantification, it also — as has been seen throughout this book — relies on the qualitative values of words, images and metaphors. This can even happen simultaneously: during the discussions that led up to the IPCC's Third Assessment Report of 2001, for example, a room full of scientists discussed for an entire week whether or not to include the three-word phrase 'discernable human influence.' Only three words, perhaps, but three extremely potent words (both qualitatively and quantitavely speaking), that between them tell a vast and potentially world-altering story.

Martin John Callanan's ongoing *Text Tends* series offers a deadpan encounter with exactly this kind of quantification of language. Using Google data the series explores the vast mine of information that is generated by the search engine's users, each animation taking the content generated by search queries and reducing the process to its essential elements:

search terms *vs.* frequency of search over time, presented in the form of a line graph.

In the online manifestation of the *Text Trends* animations — see http://greyisgood.eu/texttrends — the viewer watches as the animations plot the ebb and flow of a series of paired search terms keyed into Google over the last five years by Internet users around the world. In the case of the climate change sequence featured here, pairs of words such as: 'nature' — 'population'; 'climate' — 'risk'; 'consensus' — 'uncertainty'; 'Keeling curve' — 'hockey stick', spool out matter-of-factly, like a live market index, allowing the implied narrative content of these word comparisons (along with their accumulated cultural and emotional baggage) to play themselves out before us. In contrast to the hyper-interactivity of emerging news aggregators and information readers, *Text Trends* explores our perceptions of words presented as connotation-rich fragments of continually updated time-sensitive data.*

As an investigation into both the generation and representation of data, *Text Trends* offers a visual critique of the spectacularization of information, a cultural tic that continues to generate the endless roll of statistically compromised wallpaper that surrounds so much public science debate, and which this book — *Data Soliloquies* — has in large part been about.

* In its first sixteen months online, *Text Trends* received nearly 24,600 online views. The stills reproduced here are taken from a new animation for the UCL Environment Institute, Spring 2009. The first and last stills are from the first animation.

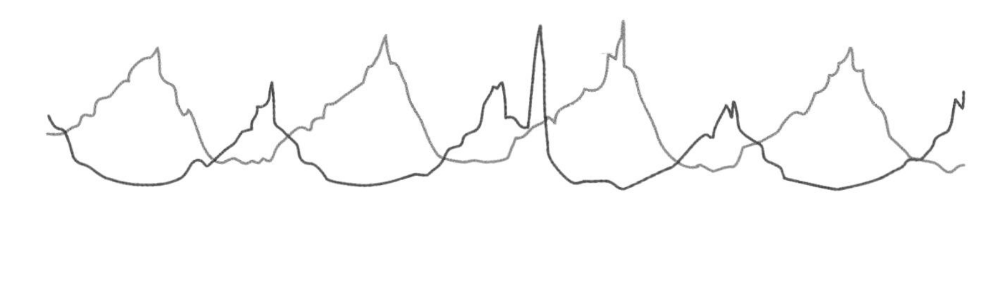
summer
winter
2004
2005
2006
2007

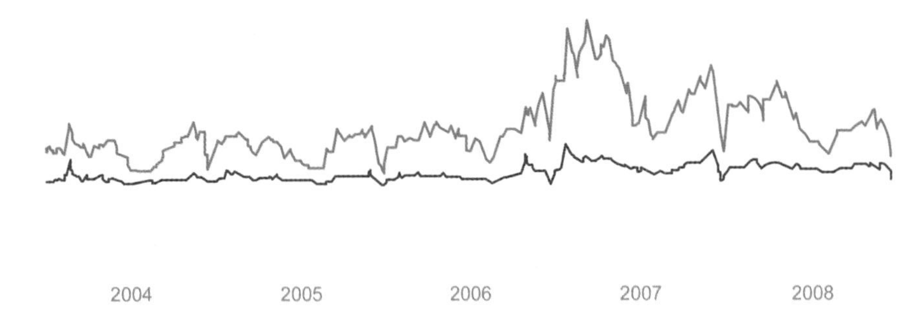
global warming
climate change
2004
2005
2006
2007
2008

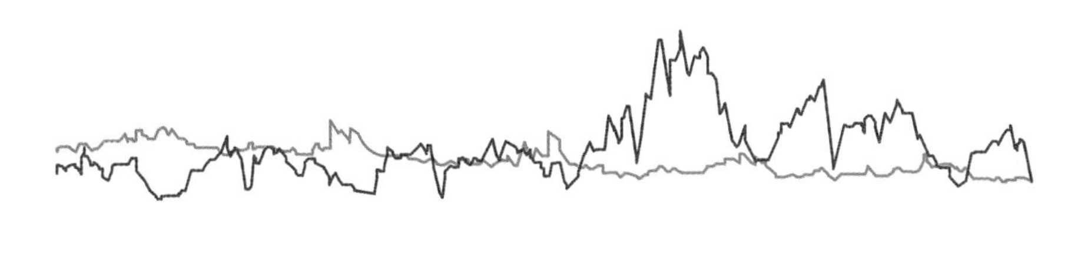
cooling
warming
2004
2005
2006
2007
2008

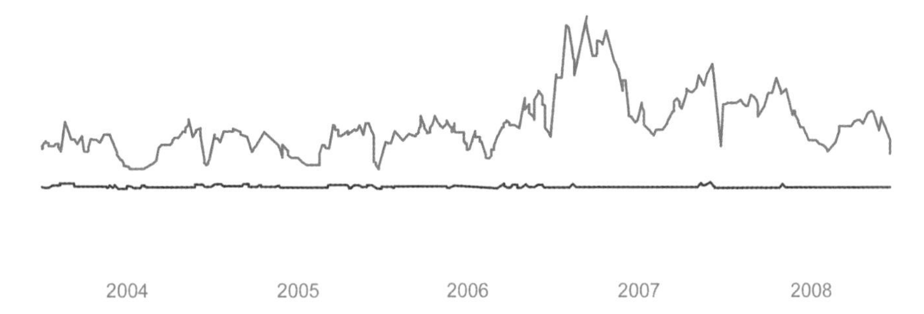
global warming
trade deficit
2004
2005
2006
2007
2008

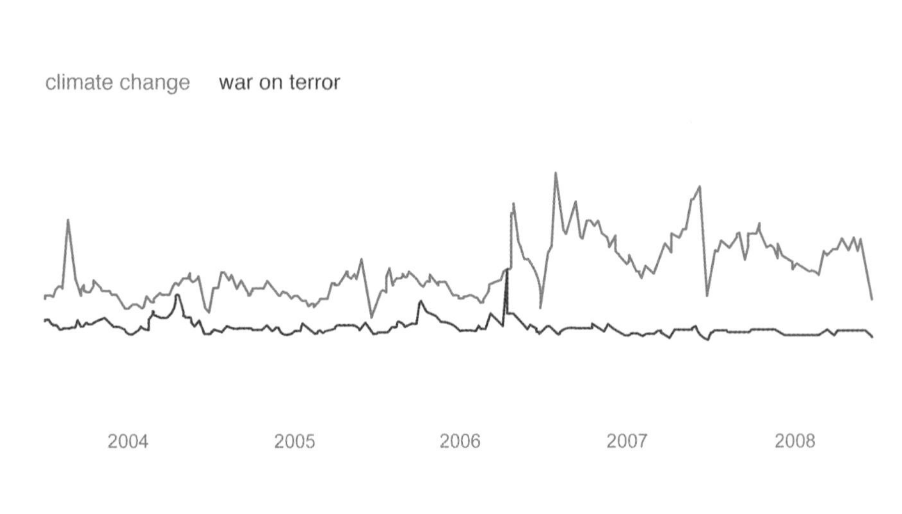
climate change
war on terror
2004
2005
2006
2007
2008

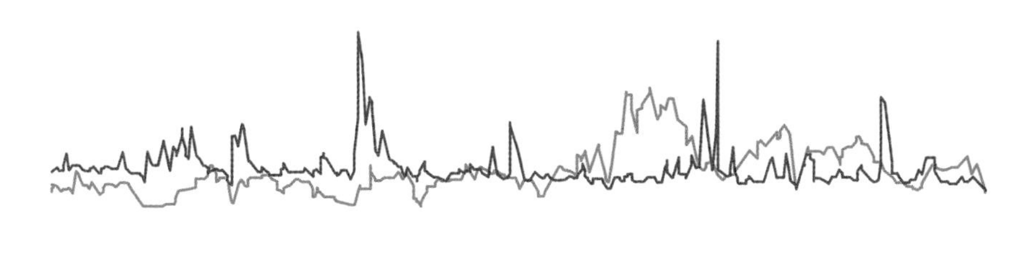
global warming
flood
2004
2005
2006
2007
2008

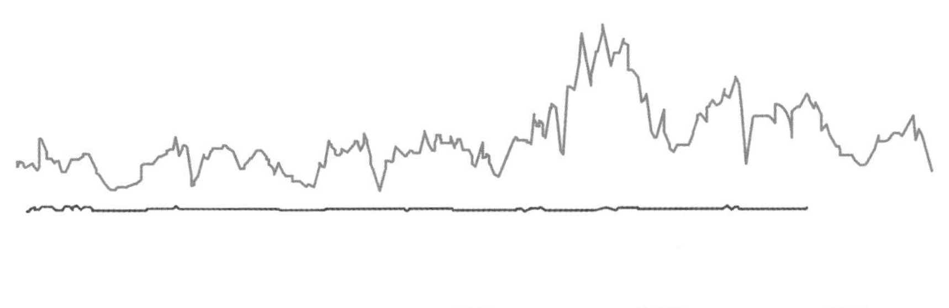
global warming
violence on tv
2004
2005
2006
2007
2008

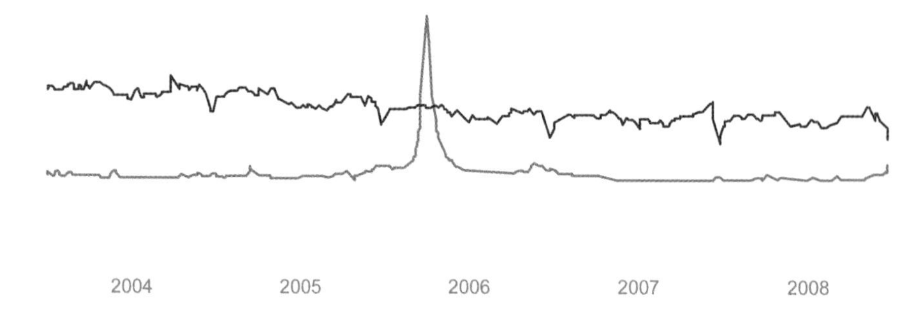
ice age
risk
2004
2005
2006
2007
2008

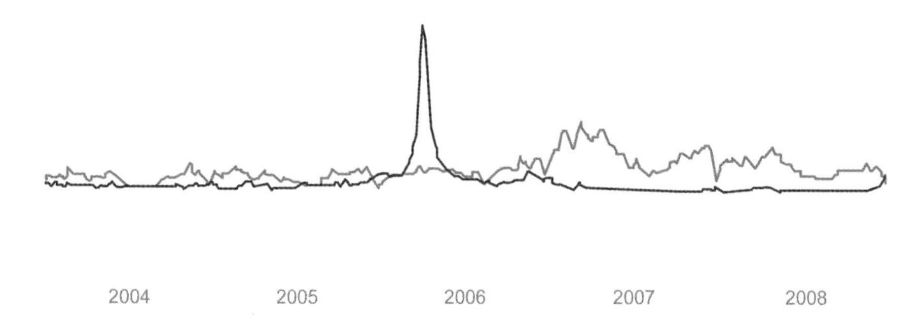
global warming
ice age
2004
2005
2006
2007
2008

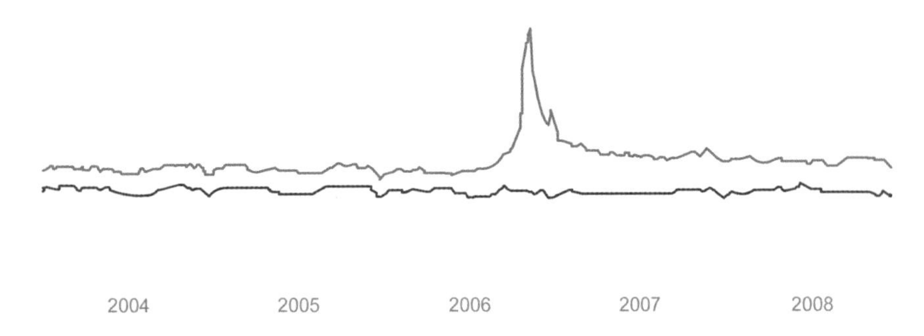
carbon
hydrogen
2004
2005
2006
2007
2008

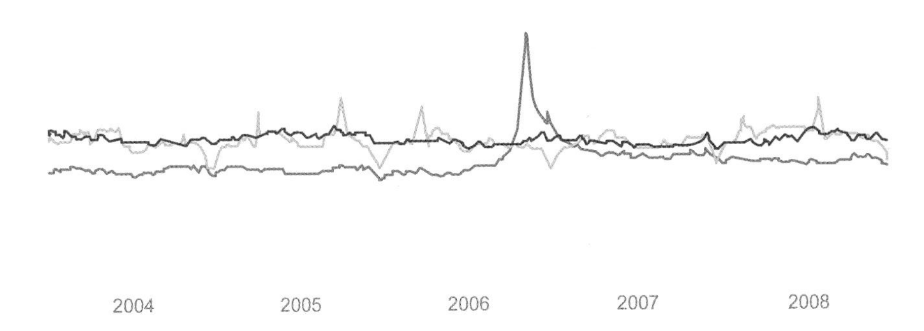
carbon
wind
solar
2004
2005
2006
2007
2008

ozone hole gm food

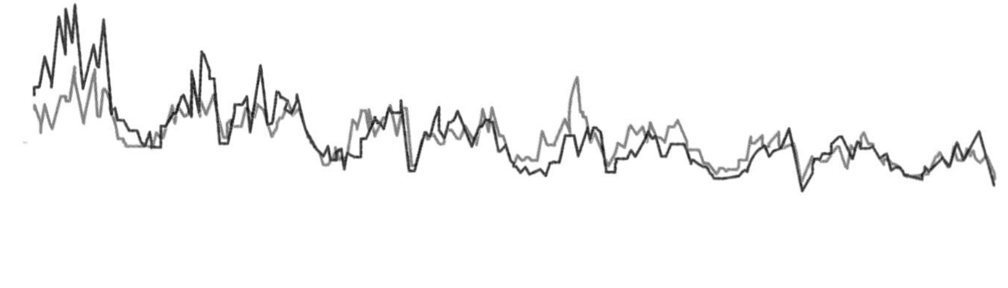

2004 2005 2006 2007 2008

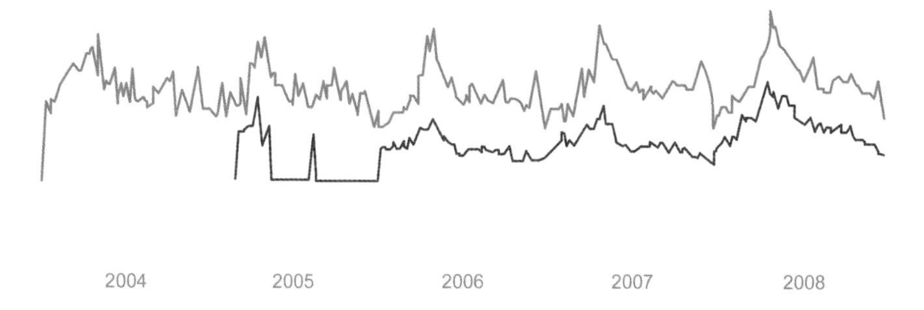
plant a tree
grow vegetables
2004
2005
2006
2007
2008

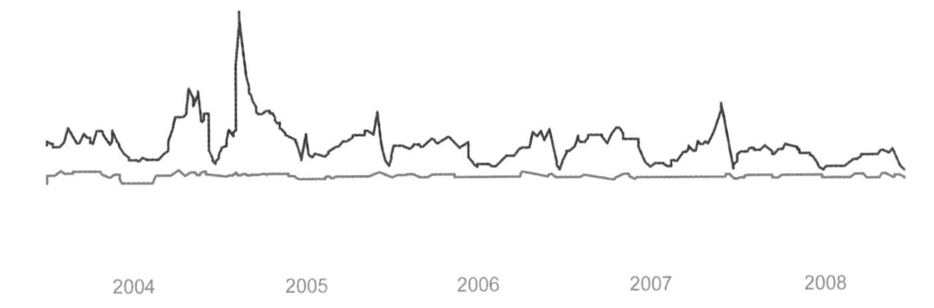
montreal protocol
kyoto protocol
2004
2005
2006
2007
2008

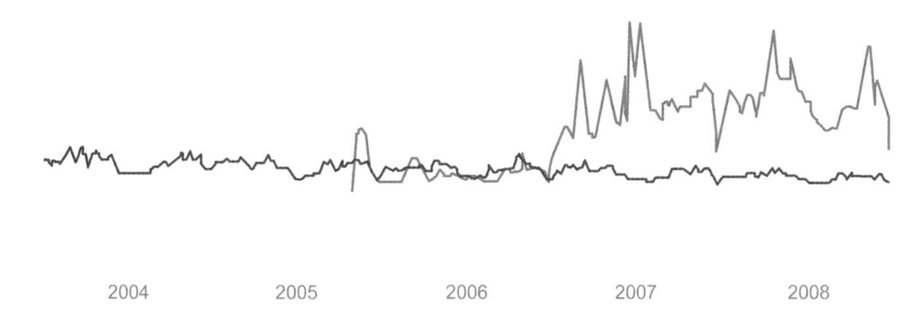
carbon footprint
ozone hole
2004
2005
2006
2007
2008

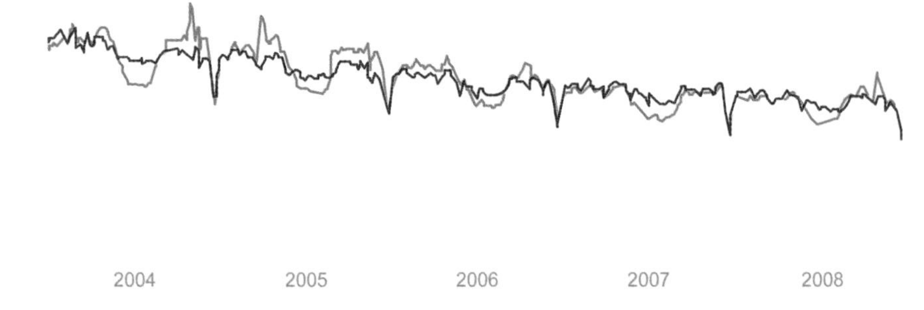
population
resources
2004
2005
2006
2007
2008

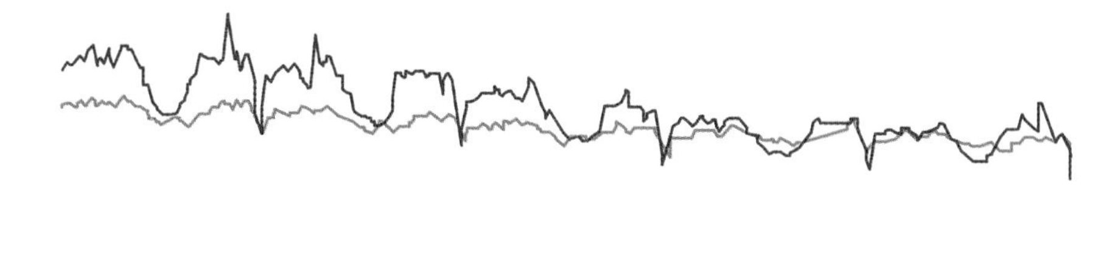
nature
population
2004
2005
2006
2007
2008

keeling
tufte
2004
2005
2006
2007
2008

keeling curve
hockey stick
2004
2005
2006
2007
2008

global
risk

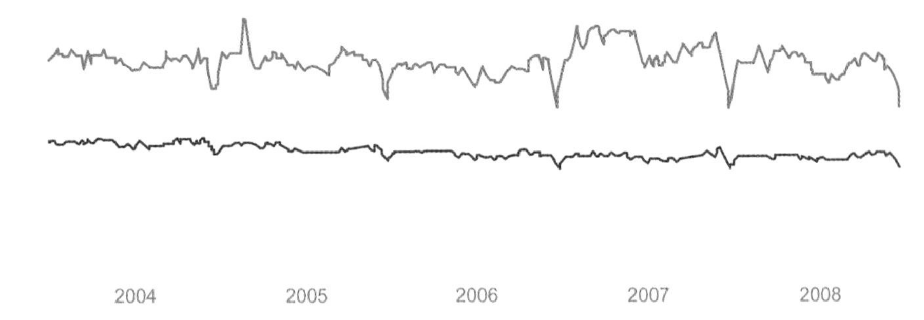
2004
2005
2006
2007
2008

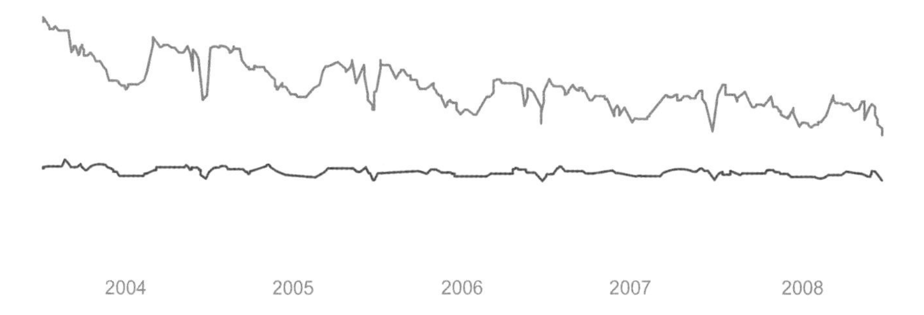
science
climate
2004
2005
2006
2007
2008

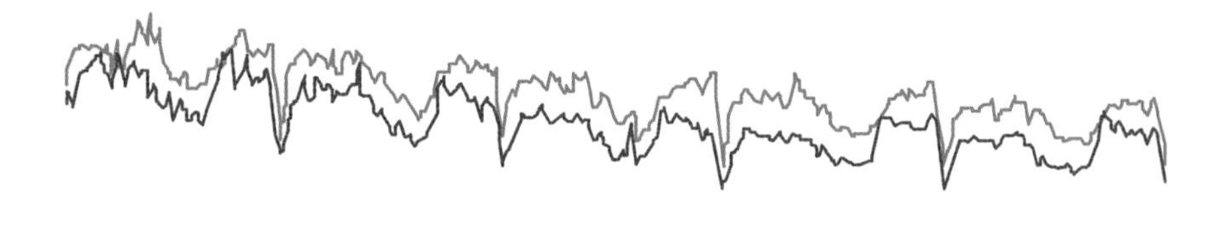
consensus
uncertainty
2004
2005
2006
2007
2008

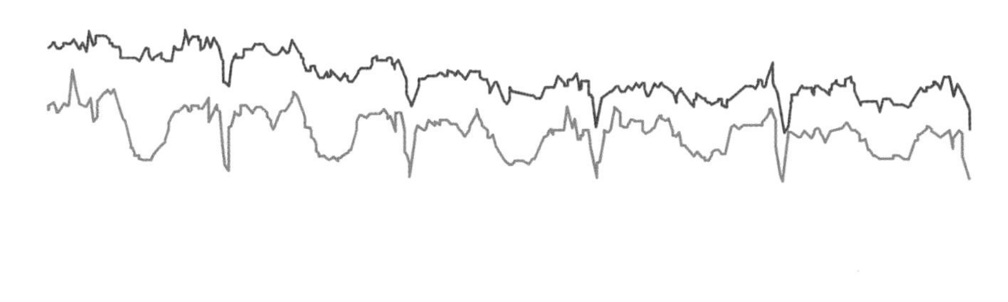
climate
risk
2004
2005
2006
2007
2008

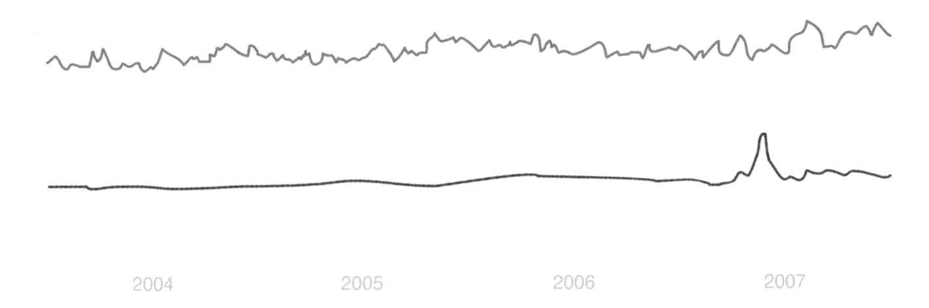
now
later
2004
2005
2006
2007

Books must follow sciences, and not sciences books.

Francis Bacon, 1616

Adams, Jon, *Interference Patterns: Literary Study, Scientific Knowledge, and Disciplinary Autonomy* (Lewisburg, Pa.: Bucknell University Press, 2007)

Best, Joel, *Stat-Spotting: A Field Guide to Identifying Dubious Data* (Berkeley, Ca.: University of California Press, 2008)

Callanan, Martin John, *Letters 2004-2006: Confirmation That You Still Exist; I Respect Your Authority; When Will It End; One London* (London: Book Works, 2007)

Chandler, James, *et al* (eds), *Questions of Evidence: Proof, Practice, and Persuasion across the Disciplines* (Chicago: Chicago University Press, 1994)

Cox, Robert, *Environmental Communication and the Public Sphere*, 2nd edn (Thousand Oaks, Ca.: Sage Publications, 2009)

Daston, Lorraine and Peter Galison, *Objectivity* (New York: Zone Books, 2007)

Desrosières, Alain, *The Politics of Large Numbers: A History of Statistical Reasoning*, trans. Camille Naish (Cambridge, Mass.: Harvard University Press, 2002)

Gregory, Jane, and Steven Miller, *Science in Public: Communication, Culture, and Credibility* (New York: Perseus Books, 1998)

Hamblyn, Richard, *The Cloud Book: How to Understand the Skies* (Newton Abbot: David & Charles/Met Office, 2008)

—, 'The whistleblower and the canary: rhetorical constructions of climate change', *Journal of Historical Geography* 35 (2009), 223-36

—, *Extraordinary Clouds: Skies of the Unexpected from the Beautiful to the Bizarre* (Newton Abbot: David & Charles/Met Office, 2009)

—, *Terra: Tales of the Earth* (London: Picador, 2009)

King, David A., 'Climate change science: adapt, mitigate, or ignore?', *Science* 303 (2004), 176-77

Maslin, Mark, *Global Warming: A Very Short Introduction*, 2nd edn (Oxford: OUP, 2008)

—, J. Corfee-Morlot *et al*, 'Global warming in the public sphere', *Philosophical Transactions of the Royal Society, Series A* 365 (2007), 2741-2776

Nelkin, Dorothy, *Selling Science: How the Press Covers Science and Technology*, 2nd edn (New York: W. H. Freeman, 1995)

Pauwels, Luc (ed.), *Visual Cultures of Science: Rethinking Representational Practices in Knowledge Building and Science Communication* (Hanover, NH: University Press of New England, 2006)

Porter, Theodore M., *Trust in Numbers: the Pursuit of Objectivity in Science and Public Life* (Princeton, NJ: Princeton University Press, 1995)

Sagan, Carl, *Billions & Billions: Thoughts on Life and Death at the Brink of the Millennium* (New York: Random House, 1997)

Shapin, Steven, *A Social History of Truth* (Chicago: Chicago University Press, 1994)

Tufte, Edward R., *The Visual Display of Quantitative Information* (Cheshire, Conn.: Graphics Press, 1983)

—, *Envisioning Information* (Cheshire, Conn.: Graphics Press, 1990)

—, *Visual Explanations: Images and Quantities, Evidence and Narrative* (Cheshire, Conn.: Graphics Press, 1997)

—, *Beautiful Evidence* (Cheshire, Conn.: Graphics Press, 2006)

Weart, Spencer, *The Discovery of Global Warming*, 2nd edn (Cambridge, Mass.: Harvard University Press, 2008)

ACKNOWLEDGEMENTS

We are particularly grateful to Mark Maslin, and Marianne Knight, Director and Deputy Director of the UCL Environment Institute, for their unstinting support over the course of our residencies and beyond.

We would also like to thank the other members of the Environment Institute for their friendly collegiality, especially Sarah Bell, Julien Harou, Phil Hopkins, Yvonne Rydin, and Simonetta Tunesi; our thanks are also due to Susan Irvine, Mark Ford, Gregory Dart, Philip Horne and John Mullan of the UCL English department; Chris Cornish and Tom Lomax of the Slade Centre for Material Research, UCL; Martin Watmough and Gregor Anderson of the Digital Manufacturing Centre, Bartlett School of Architecture, UCL; Andy Hudson-Smith, Maurizio Gibin and Alex Singleton at UCL Centre for Advanced Spatial Analysis; John Aiken, Susan Collins, Zara Dinnen, Michael Duffy, Simon Faithfull, Mick Farrell, Dryden Goodwin, Tim Head, Klaas Hoek, James Keith, Sharon Morris, Caroline Nicholas, Sarah Pickering, Joy Sleeman and Jon Thomson of the Slade School of Fine Art, UCL; Nicholas Alfrey, Stephen Daniels, and Julie Sanders of the Water, Culture and Society project, University of Nottingham; Greg J. Smith of Serial Consign; Liz Lawes, Simon Gould, Sussanah Chan and Fiona Davidson of UCL; Neil Lonie, The University of Dundee; Dan Pisut of National Oceanic and Atmospheric Administration; Barry Gromett and Sarah Tempest, The Met Office; and Nicola Triscott and Rob La Frenais of Arts Catalyst; and the nameless individuals at: The European Space Agency R&D Section; European Organisation for the Exploitation of Meteorological Satellites; Mesoscale Dynamics and Modeling Group, Goddard Space Flight Center, NASA.

Plus, special thanks to Jon Adams of the London School of Economics for inadvertently coming up with the title.

Earlier versions of these chapters have been presented as papers at a variety of conferences and seminars, and we would like to thank the organizers of: the

'Narratives of Climate Change' plenary session at the Institute of British Geographers/RGS Conference, 31 August 2006; 'Weathering Climate Change' at the University of Hull, 10 October 2007; the Arts Catalyst 'Eye of the Storm' conference, Tate Britain, 19-20 July 2009; and 'RETHINK — Contemporary Art and Climate Change' (part of the cultural programme for the COP15 meeting in Copenhagen, December 2009): see http://www.rethinkclimate.org.

PICTURE CREDITS

Cover image, pages 64-65, 68-73, 79-101: Martin John Callanan; 24: Gerald Meehl, National Center for Atmospheric Research; 25: NOAA/ESRL; 26: redrawn from a screengrab of *An Inconvenient Truth* © 2006 Paramount Pictures; 27: IPCC/Mann *et al* 1998; 30, 31: NASA/JPL; 33: NASA/Goddard Institute for Space Studies; 34: US Global Change Research Program; 39: Stephen Schneider and Richard Moss; 40: NASA/Goddard Space Flight Center Scientific Visualization Studio; 49: Photo by RH, used with kind permission of Tesco plc; 61: NASA; 63: Veres Viktor/NASA.

These images have been derived from many sources and acknowledgement has been made wherever possible. If any image has been used without due credit or acknowledgement, through no fault of our own, apologies are offered. If notified, the publisher will be pleased to rectify any errors or omissions.

INDEX

Richard Hamblyn is an environmental writer and historian; his books include *Terra: Tales of the Earth*, a study of natural disasters; *The Invention of Clouds*, which won the 2002 Los Angeles Times Book Prize; *The Cloud Book* and *Extraordinary Clouds* (both in association with the Met Office). He is currently editing *The Picador Book of Science*, and researching a book about man-made landscapes.

Martin John Callanan is an artist whose work spans numerous media and engages both emerging and commonplace technology. His work includes translating active communication data into music; freezing in time the earth's water system; writing thousands of letters; capturing newspapers from around the world as they are published; taming wind onto the Internet and broadcasting his precise physical location live for over two years. Martin is currently Teaching Fellow in Fine Art Media at the Slade School of Fine Art in London.

http://greyisgood.eu

The UCL Environment Institute was established in November 2003 as a focus for interdisciplinary environmental research across UCL, as well as to improve links between those who carry out environmental research, and those with need of its findings, notably policy makers and other public and private sector interests.

http://www.ucl.ac.uk/environment-institute